THE SCIENCE
of JAMES
SMITHSON

———+———

THE SCIENCE
of JAMES
SMITHSON

———+———

DISCOVERIES
from the
SMITHSONIAN FOUNDER

———+———

STEVEN TURNER

Smithsonian Books
WASHINGTON, DC

Funding for this book was provided in part by Smithsonian Institution Scholarly Press

Published by Smithsonian Books
Director: Carolyn Gleason
Senior Editor: Jaime Schwender
Assistant Editor: Julie Huggins

Edited by Gregory McNamee
Cover design by Pete Garceau
Designed by Gary Tooth

This book may be purchased for educational, business, or sales promotional use. For information, please write: Special Markets Department, Smithsonian Books, P.O. Box 37012, MRC 513, Washington, DC 20013

Library of Congress Cataloging-in-Publication Data
Names: Turner, Steven, author.
Title: The science of James Smithson : discoveries from the Smithsonian
    founder / Steven Turner.
Description: Washington, DC : Smithsonian Books, [2020] | Includes bib-
    liographical references and index. | Summary: "An exploration of the
    scientific career of James Smithson, who left his fortune to establish
    the Smithsonian Institution"—Provided by publisher.
Identifiers: LCCN 2020017952 (print) | LCCN 2020017953 (ebook) |
    ISBN 9781588346902 (hardcover) | ISBN 9781588346933 (epub)
Subjects: LCSH: Smithson, James, 1765-1829. | Smithsonian Institution—
    History. | Scientists—Great Britain—Biography.
Classification: LCC Q143.S6 T87 2020 (print) | LCC Q143.S6 (ebook) |
    DDC 509.2 [B]—dc23
LC record available at https://lccn.loc.gov/2020017952
LC ebook record available at https://lccn.loc.gov/2020017953

Printed in the United States of America
24 23 22 21 20   1 2 3 4 5

For permission to reproduce illustrations appearing in this book, please correspond directly with the owners of the works. Smithsonian Books does not retain reproduction rights for these images individually or maintain a file of addresses for sources.

For Ikuko

CONTENTS

INTRODUCTION

WHO WAS JAMES SMITHSON? That question has been asked since the 1830s, when a little-known Englishman left his fortune, as his will specified, to "the United States of America, to found at Washington, under the name of the Smithsonian Institution, an establishment for the increase and diffusion of knowledge among men."

This was not Smithson's first choice. He initially left the estate to his nephew, who controlled it until that nephew died in 1835, after which it was then supposed to go to his heirs. But the nephew had been childless, and in that case the will, in its very last clause, required the money go to the United States.

Smithson had not talked to anyone about this, and the phrase "an establishment for the increase and diffusion of knowledge among men" served as the only instructions he ever gave for how the money should be used. In 1838, when Smithson's estate arrived in New York harbor, it consisted of his clothes and personal possessions, his papers, scientific instruments, and extensive mineral collection, and several boxes of gold coins. The coins were promptly melted down and reminted into $508,318.46 in US currency—about 1.5 percent of the entire federal budget at the time. It was an enormous sum, and it took Congress almost eight years to decide what to do with it. Eventually Congress decided to honor Smithson's request, and in 1846 the funds were used to found the Smithsonian Institution, the vast museum complex whose buildings now line the National Mall in Washington, DC.

The Smithsonian thrived and grew, but there was always a desire within the institution to learn more about its founder. Smithson was known to have been a man of science, but beyond that little was known about him. The Smithsonian Annual Report for 1857 announced that a "list of the papers published by Smithson, and a record of all the facts which could be gathered in relation to him, have been made, to serve hereafter for a more definite account of his life and labors than has yet appeared." That project was not yet completed when a fire that ravaged the Smithsonian Castle in 1865 destroyed most of Smithson's papers and all of his artifacts—including his huge mineral collection.

The Smithsonian Castle was rebuilt, and in 1879, to commemorate the fiftieth anniversary of Smithson's death, the institution made another attempt to reconstruct its founder's history. Requests for information were posted in major American newspapers and journals and privately circulated in England.

Around the same time, an effort was made to understand Smithson's science and identify his scientific articles. Twenty-seven published scientific articles were found and subsequently reprinted in *The Scientific Writings of James Smithson* (1879). Along with the articles, this commemorative volume also included two reviews of Smithson's "scientific character." One had been written by American chemist Walter R. Johnson in 1844, and the other was a newly commissioned work by twenty-five-year-old American mineralogist John Robin McDaniel Irby. Johnson's review was primarily factual, but Irby's expressed an opinion that would become the standard assessment of Smithson for the next century. He wrote:

> We could wish Smithson's name to have been coupled with some great discovery, or with the apprehension of some far-reaching law that would have formed a worthy inscription for the portal of his institution. Though this be not gratified, we shall find that he appreciated the great problems before him and attempted their solution; that he knocked earnestly and worthily at the portal of great knowledge, and that although it was denied him to be the first to enter into the greater chambers, he was, nevertheless, no unworthy seeker.

This conclusion and its unspoken sense of disappointment set the tone for all subsequent assessments of Smithson's work. Even within the Smithsonian, the view was taken that although Smithson had been dedicated and hardworking, he had been little more than a scientific dabbler or a second-tier figure.

In 1904, Samuel Langley, the Smithsonian secretary, wrote in the *Scientific American Supplement* that Smithson "is not to be classed among the leaders of scientific thought; but his ability and the usefulness of his contributions

THE SMITHSONIAN'S FOUNDER.
From Nature.

We would earnestly draw the attention of our readers to the fact that the Secretary of the Smithsonian Institution, Washington, United States, of which Mr. James Smithson was the founder, is desirous of obtaining information respecting that gentleman to assist in the preparation of a memoir. James Smithson, F. R. S., was the son of Hugh. first Duke of Northumberland, and Elizabeth, heiress of the Hungerfords of Audley, and niece of Charles, Duke of Somerset. In 1826 he resided at Bentinck-street, Cavendish-square. He · died in 1829. The following are some of the points on which information is desired: "John Fitall, a trusted servant of Mr. James Smithson, died June 14, 1834, at Bush House, Wanstead, Essex, England. Have his heirs any relics or mementoes of Mr. Smithson—any notes, letters, &c.? Mr. Charles Drummond, a London banker, was the Executor of Mr. Smithson. Can we procure originals or copies of any letters of Mr. Smithson from him? What do the records of the Royal Society say as to the election of James Lewis Macie as a Fellow? Perhaps a report was made to the Council as to his qualifications. What can be learned of the disagreement between Mr. Smithson and the Council of the Royal Society? Mr. Wheatstone knew of it. Do any of the surviving members remember the circumstances? Information relative to Henry Louis Dickinson, (half-brother of James Smithson,) Colonel of the Eighty-fourth Regiment of Foot. Information relative to the college life of James Lewis Macie, a graduate of May 26, 1786, of Pembroke College, Oxford University. Letters from James Smithson to Sir Humphrey Davy, Sir Davies Gilbert, the Hon. Henry Cavendish, Dr. W. H. Wollaston, Mr. Smithson Tennant, Dr. Joseph Black, Dr. Hutton, M. Arago, M. Gay Lussac, M. Cordier, M. Haüy; M. Klaproth, M. A. C. Becquerel, M. Fanjas De St. Fond, Mr. Thornton, Mr. Maclaire, Mr. William Thomson, or any original letters of Mr. Smithson. Can the original manuscripts be found of Mr. Smithson's communications to the Royal Society or to Thomson's "Annals of Philosophy?" Can Mr. Smithson's authorship of papers or articles in any scientific journals be identified? What can be learned of Mr. Smithson's mother, Mrs. Macie, or of Col. Henry Louis Dickinson's mother, Mrs. Mary Ann Coates? At what number in Bentinck-street did Mr. Smithson reside? (He held apartments, was not a householder,) Had he at any time any other residence; if so, where?" Any information on the above points should be addressed to Prof. Spencer F. Baird, care of William Wesley, No. 28 Essex-street, Strand, London, the agent of the Smithsonian Institution.

This request for information about Smithson appeared in the *New York Times* on June 27, 1880, after appearing in a recent issue of the journal *Nature*. A similar request for information was also privately circulated in England. The New York Times Archive.

to knowledge, cannot be doubted." Twenty-five years later, on the centenary of Smithson's death, no one in the Smithsonian objected when the respected journal *Science* concluded that "Smithson missed being a great scientist as a research worker, but he was a thorough and an indefatigable one." Even as late as 1965, in a book supported by the Smithsonian and partially written by its secretary, this view was firmly embraced: "Smithson's intellectual curiosity was endless . . . [however] the very catholicity of his view kept him from concentrating on any particular field. But his example served to enhance his reputation in his own time, and his studies gave hints which were useful to future research chemists."

The assumption underlying all these statements, that the growth of scientific understanding depends on the extraordinary insights of a few "great" individuals, has since fallen out of favor. Historians of science no longer try to identify each age's geniuses but rather look to the role of institutions, networks, tools, and scientific debate in studying the scientific process. This suggests that Smithson deserves more credit, because these were all areas in which he was active. Even if this were not the case, the Smithsonian's original assessment of its founder would be suspect, because it doesn't appear that the people who made that assessment understood his work.

To give one example, *The Scientific Writings of James Smithson*, the book the Smithsonian published as Smithson's collected works, reprinted twenty-seven articles. But not all of them are actually articles, and not all were written by Smithson; the fourth "article" is actually a review of the fifth article, written by the chemist Humphry Davy. To be fair, Smithson's articles are difficult to understand. His writing style was terse, and most of the topics he wrote about are not currently studied. On top of that, the tools and methods he used are unfamiliar, as are the names he used for them. It is perfectly understandable that a study of Smithson's science made 140 years ago should have weaknesses and seem dated. The history of science was not a separate discipline when that book was written, and the insights and scholarly study of the ensuing years were not available to its writers. They did the best they could, but the time has come to take a fresh look at Smithson's science, and that is what this book hopes to accomplish.

Almost all we know about Smithson's science comes from his twenty-six published articles, and so these became my starting point. I present them here in terms of their own scientific, social, and historical context. These stories describe several new and unexpected aspects of Smithson's work, and, taken as a whole, they provide some surprising insights into Smithson's personality and intellectual life. But there is an aspect of Smithson that can only be appreciated by looking closely at the material side of his scientific work: his tools and scientific practices.

Smithson was an accomplished analytical chemist, and much of his importance lies in the tools and scientific methods he developed. These, too, have their own history, and only after I began repeating Smithson's experiments did I truly begin to understand how important that history is. Following his descriptions step-by-step, using the same tools and materials, seeing, smelling, and even hearing the same things he did, all this made his articles come alive for me. Replicating Smithson's experiments was like opening a window into his time, and one of the goals of this book has been to communicate that sense of discovery to the reader.

In telling this story, I wanted to present Smithson's science as he would have known it, so I made little attempt to link what he did to modern science or modern understanding. As much as possible, I have tried to preserve his terminology and spelling, with the notable exception of Smithson's name. Born James Louis Macie, he used that name until he was in his thirties, when he changed it to James Louis Smithson (Smithson being his father's original name). To avoid confusion, "Smithson" has been used throughout the book. I should also note that I have resisted calling Smithson or his colleagues "scientists." Although "science" was widely used in his time, "scientist" was not. Smithson and his contemporaries were "natural philosophers" and "savants."

. . . .

A project like this book is a group effort, and there are many people without whom it could not have happened. My own interest in Smithson grew out of a series of conversations with Heather Ewing, whom I met while she was writing her excellent biography *The Lost World of James Smithson: Science, Revolution, and the Birth of the Smithsonian* (Bloomsbury, 2007). Through a combination of diligent research and scholarship, she reconstructed Smithson's personal life in remarkable detail. Although Smithson's science was not the focus of her work, she argued that the traditional view of it was incomplete. With her encouragement, I began this study.

Understanding Smithson's work would not have been possible without the encouragement and special knowledge of others. Daniel Kelm first introduced me to the wonders of eighteenth-century chemistry and the possibility of recapturing the spirit of discovery behind Smithson's articles. The term "volunteer" fails to capture how important Jeff Gorman and Frank Cole were to me; colleagues, friends, and traveling companions (along with Ginni and Mary Lou) are words closer to the truth. The same applies to Nathan Karch, who helped me in critical ways later in the project. My deepest thanks to them all.

As a curator at the Smithsonian Institution, I had access to a network of colleagues and support that never ceased to amaze. In particular, David Allison provided critical advice in framing the scope of this book and the questions that it tries to answer. Pam Henson gave sage advice when I needed it most and always served as a role model. David DeVorkin served as another kind of role model and friend, as did Michelle Delaney and Jane Milosch. The Smithsonian Institution Libraries provided resources that were invaluable to me, including Nancy Gwinn, Leslie Overstreet, Lilla Vekerdy, and Jim Roan. Finally, my thanks to Frank James of University College London, who not only gave good advice but also inspired me to think originally.

1.

THE LONG ROAD
TO STAFFA

JAMES SMITHSON (1765–1829) was born into wealth and privilege. His mother was Elizabeth Hungerford Macie, descended from a proud, distinguished, and still wealthy family; his father was Hugh Percy (formerly Hugh Smithson), the Duke of Northumberland, an influential advisor to the king and one of the richest men in England. But she was a widow, and he was married, and when she became pregnant these facts suddenly mattered very much. They mattered so much that she quickly left England and moved to Paris, where she lived in seclusion until she gave birth. To this day we do not know when or exactly where James Smithson was born.

We also know very little about his childhood, but it generally seems to have been turbulent, with his mother tumbling from one personal crisis into another. His physical needs were cared for, but there was little of what might be called a normal childhood. He seems to have been raised mostly by surrogates, and when he was eight, he was put in the care of a guardian, one of his mother's solicitors, who appears to have enrolled him in a London boarding school. This may have been a necessary step, but not one that was likely to relieve a young boy's sense of isolation.

From an early age, Smithson seems to have been highly motivated to study chemistry. The origin of that motivation has long been a mystery, but he left a clue about it in one of his scientific articles. The Archive of London's Royal Society still holds eight of Smithson's original handwritten manuscripts, and in an unpublished footnote to one of them he refers to a book: "This edition of Wilson's chemistry, [is] dedicated to my Father who was well informed in chemistry and from whose conversation I imbibed when a child my taste for it." The book was a popular textbook, George Wilson's *A Course of Practical Chemistry*, rewritten in 1746 by the chemist William Lewis and published under his own name. Lewis dedicated that edition to "Sir Hugh Smithson," with whom he apparently was developing a method to refine platinum.

This footnote is the only known evidence of any relationship between Smithson and his father, and there is little reason to think that they met more than a few times. But it is clear that this relationship was important to Smithson, and that he saw chemistry as a personal link between them. It is striking that throughout his adult life as well, science would be the topic that, more than any other, connected him to the important people in his life.

In 1782, at the age of seventeen, Smithson entered Oxford University. His wealth allowed him to enroll as a "Gentleman Commoner," the highest student rank outside the nobility, and this in turn entitled him to a variety of privileges. He received larger rooms and the right to wear black silk gowns and a black-tufted velvet cap (signifying his station), as well as holding special seats at chapel and at the dining table. It also freed him from having to work in the college, as many other students needed to do to cover their expenses. Smithson hired someone to clean his rooms, and he had a personal tutor to make sure that he passed all his examinations. Smithson was thus largely free to pursue whatever interested him at Oxford, and he applied himself diligently to the study of science. It was later reported that Smithson "had the reputation of excelling all other resident members of the University in the knowledge of chemistry."

Smithson chose Pembroke as his college, an important choice. While his classes were organized by the university, Pembroke was where he would sleep, eat, study, and entertain. The academic and social tone within each college was largely determined by the personality of its master. The master of Pembroke was William Adams, a cleric who was said to be "considerably deep in chemistry," and who encouraged and supported that interest in his charges. It is probably no accident that the four notable chemists to come out of Oxford in the later eighteenth century—one of whom was Smithson—all matriculated from Pembroke. Adams was also notably less politically conservative than most of his Oxford colleagues, a trait that seems to have characterized many of the students in Pembroke.

Smithson was fortunate to enter Oxford when he did. If he had enrolled just a few years earlier, he would have found that the chemistry laboratory was badly deteriorated and that the little chemistry instruction being offered at the university was out of date. A revolution in chemistry was under way in France, but Oxford students were still being taught the same obsolete chemical system that had been instituted more than seventy years earlier.

But in 1781, just a year before Smithson arrived, that all changed. Martin Wall, a graduate of the University of Edinburgh Medical School and an accomplished chemical experimenter, was appointed reader in chemistry. Wall was also an honorary member of the important Literary and Philosophical Society of Manchester, to which he would submit several articles. Perhaps his most important qualification, however, was that he had taken the Edinburgh course of the brilliant and pioneering chemist Joseph Black. The chemistry course that he put in place, which Smithson would take, drew heavily on that experience.

Wall also consulted Black about the long-needed renovation of the chemistry lab, located in the basement of the Ashmolean Museum, now the History of Science Museum. Large, well-lighted, and well-ventilated, with limestone walls and a high, arched limestone ceiling, it was an ideal space for chemical experiments. Wall delivered his lectures in an adjacent room, formerly used for teaching anatomy and now fitted with benches, and a third room provided storage space for chemicals, apparatus, and possibly a chemical library. Wall was able to have the existing chemical furnaces, which were original to the lab and nearly a century old, replaced with modern chemical furnaces of Black's design. He was also able to replace all the existing lab equipment, which was unceremoniously dumped behind the building. By the time Smithson arrived, Oxford had a dynamic chemistry instructor, a revised curriculum, and an up-to-date laboratory.

This upgrade of Oxford's chemistry course was justified as a necessary step to improve the university's medical degree program, but the few records that survive from this time indicate that many of the students in Wall's class were either studying divinity or, like Smithson, simply interested in science. The surprising fact is that in the late eighteenth century, the study of chemistry was extremely popular. Chemists still struggled to explain how materials combined, and even in Smithson's time there was still robust debate about things as basic as the nature of fire. But the practical value of chemistry was undeniable, and there was a widespread expectation that chemists were about to uncover some of nature's fundamental truths. This philosophical side of chemistry was of great interest to Smithson, as it was to most students with an interest in science, and it had an unexpected appeal to the divinity students. One of them explained in a letter to his father, "The business which most interrupts my study of the Greek Testament is the chemical lecture. This intricate, but entertaining science, demands a very large share of attention." Although it was "useless to enquire how far a knowledge of chemistry may promote my future prospects in life, or assist in qualifying any man for a clergyman," he gave three ways in which it might be beneficial. First, like every branch of knowledge, chemistry offered the benefit of "expanding and enlarging the mind." Second, and this was no trivial point for a man preparing for a career giving sermons, "it may furnish [a] variety of hints, allusions, expressions [for use] in every kind of composition." Finally, a familiarity with chemistry promised to arm the future clergyman against the arguments of "Voltaire and all his disciples" when they tried to use science to deny religion.

If it seems odd that medical, divinity, and science students all ended up in the same class, the explanation is simple: Wall's class was the only chemistry course that Oxford offered. There was no advanced course in chemistry and no graduate courses. Students such as Smithson who wanted to learn more would need to do so outside of the classroom.

Wall's classroom contained almost nothing that a twenty-first-century reader would associate with the study of chemistry. There was no chart of the periodic table, no blackboard, no chemical formulas, no sinks, and no Bunsen burners—none of these things existed yet. Instead, the class consisted of Wall reading his lecture while the students followed along from a printed syllabus. Taking notes was a challenge, since the plain wood benches they sat on made using quill pens and bottles of ink impractical, and Wall spoke so quickly that even making brief notations in pencil was difficult. The students mostly wrote their notes from memory after the class ended.

In his lectures, in addition to describing the various chemical materials, methods, and equipment, Wall discussed the various competing chemical theories and their relative merits. Chemistry was still in an early stage, and even basic concepts like the definition of an element were under discussion. He also occasionally inserted a "spice of divinity" into his lectures by making observations about chemical references in the Bible. On one occasion he discussed the various methods that Moses might have used to turn the Golden Calf into powder, and on another the nature of the compound called "nitre," which is described in Proverbs.

Two other notable scientific figures instructed Smithson during his time at Oxford. The first was Thomas Hornsby, who, among his many other appointments, was the professor of natural philosophy. He taught the course on "Experimental Philosophy," which was essentially a physics course, although there was considerable overlap between chemistry and physics at this time. Smithson almost certainly took Hornsby's class, probably after completing Wall's chemistry course, which was the recommended sequence. Hornsby was noted for his extensive use of classroom demonstrations, which he used with good effect to illustrate the major points of his lectures. Several descriptions of Smithson as a young man allude to his familiarity with physics, and he probably had Hornsby to thank for this. Hornsby also offered a special course of lectures on the recently discovered gases, or "airs," which Smithson also likely took. Gases were an important new field in chemistry, and Hornsby was reported to have an elaborate new demonstration apparatus that could produce all the known gases and demonstrate their properties.

The other important scientific figure during Smithson's time in Oxford was his friend and mentor William Thomson. Thomson graduated from Oxford in 1780, just as Martin Wall was returning from his studies at the University of Edinburgh, and with Wall's encouragement he decided to study medicine at Edinburgh, which included taking the chemistry course of Joseph Black. He also took the natural history course of John Walker, which inspired him to study mineralogy. Walker's mineral collection included a number of large crystal specimens that he collected some years earlier on a trip through northwest Scotland. Thomson was fascinated by them and decided to make a collection of his own. In the summer of 1782, after completing his studies in Edinburgh and before returning to Oxford, he spent several months retracing Walker's trip and building a mineral collection.

Thomson's ambition was to teach, but back in Oxford he found himself in the not unusual position of having to wait two years for an appointment. Like many in that situation, he supported himself in the interim by taking on students for private study. Thomson used the specimens he had collected in Scotland to develop a course of lectures on mineralogy, and one of his first students was young James Smithson. The highlight of Thomson's collection was a spectacular set of large, perfectly formed white cubical crystals on a matrix of jet-black lava. The crystals were "cubical zeolites" (now called chabazites) that he found in a cave on a small, unusual island in the Inner Hebrides called Staffa.

Studying with Thomson is probably where Smithson developed his interest in mineralogy, and Thomson was amazed at his progress. A year later he reported that Smithson "dedicates his whole time to Mineralogy" and that "his proficiency is . . . already much beyond what I have been able to attain to." This was high praise, particularly considering that in 1785 Thomson would become Oxford's mineralogy lecturer. Thomson soon made arrangements for Smithson to make his own mineral collecting trip to Scotland, including a visit to Staffa.

Thomson learned that a noted French naturalist, Barthélemy Faujas de Saint-Fond, was planning an expedition to northwest Scotland, and he worked hard to place Smithson in his party. Faujas, an assistant naturalist at the Muséum National d'Histoire Naturelle in Paris, was interested in both geology and mineralogy. He already had a substantial reputation, particularly for his work on volcanoes, and now he was preparing to undertake the fieldwork for a book about the geology of Scotland. His plan was to travel from London to Edinburgh, then across Scotland to Staffa, and back across northern Scotland to Edinburgh. Along the way he would make natural history observations, collect mineral samples, and visit other points of interest. The book itself would be a mixture of travelogue and scientific treatise, a literary form that was popular at the time, when travel to distant locations was difficult and expensive.

Faujas had been in London for about a month when Smithson and Thomson met him. Faujas had not wanted for entertainment in London, since French savants of his status were not commonly encountered in England, and he was traveling with the charming Count Paolo Andreani, the first Italian balloonist. Andreani's ascent earlier in the year had been witnessed by twenty thousand of his countrymen, and his presence in the upcoming expedition lent it a distinct air of celebrity.

Members of the Royal Society had gone out of their way to entertain the pair, and on the night Smithson first met him, Faujas and Andreani were the guests of honor at the society's Thursday night meeting. Thomson arranged for him and Smithson to attend the meeting, and afterward Faujas and Smithson had a brief conversation. Faujas would later recall, "M. de Mecies [Smithson], of London, had been introduced to us a few days before our departure from London, by Mr. Thompson [sic], a very good naturalist, as a studious young man, who was much attached to mineralogy; we admitted him, with pleasure, into our party."

To a modern reader, this may suggest that he was taking Smithson along as an intern, but in the eighteenth century this practice did not yet exist. Much more likely is that Smithson was allowed to join the group because he was wealthy enough to help with the expedition's expenses and Faujas judged him to be unlikely to get in the way.

The fourth member of the expedition was William Thornton, a twenty-five-year-old medical student with broad interests who had been invited primarily for his skills as a draftsman, enabling him to produce the necessary illustrations for Faujas's book. They were also accompanied by four servants: two for Andreani and one each for Faujas and Smithson.

Faujas's decision to see Staffa was almost certainly inspired by Joseph Banks's enraptured account of his own visit to the island twelve years earlier. Banks was the handsome and adventurous naturalist who had recently sailed around the world with Captain James Cook, and the stories and collections he brought back to London from the South Pacific made him famous. In 1772 he made a second voyage, this time leading it himself, to explore the volcanic island of Iceland. Neither Banks nor any of his colleagues had any knowledge of Staffa, but as they sailed up the western coast of Scotland they were forced to anchor near the island of Mull for a few days to wait out a storm. Banks went ashore and was having breakfast with one of the island's residents when he heard about a strange nearby island made of stone columns.

Intrigued, he decided to visit the island immediately. He and his colleagues quickly returned to the ship, gathered provisions into a boat, and set out that afternoon. They had less than thirty miles of water to cross, but the wind died and they had to row most of the way, so it was nine in the evening and nearly dark by the time they finally arrived. They set up their tents, cooked dinner, and excitedly waited for dawn. Banks wrote in his journal, "Every one was up and in motion before the break of day, and with the first light arrived at the S.W. part of the island, the seat of the most remarkable pillars; where we no sooner arrived than we were struck with a scene of magnificence which exceeded our expectations, though formed, as we thought, upon the most sanguine foundations. The whole of that end of the island supported by ranges of natural pillars, mostly above 50 feet high, standing in natural colonnades."

Banks wrote about the deep sense of wonder and humility he felt as his crew worked their way along the shore and into the opening of the great sea cave: "Compared to this what are the cathedrals or the palaces built by men! mere models or playthings, imitations as diminutive as his works will always be when compared to those of nature. Where is now the boast of the architect!" Staffa would later become famous for its ability to inspire these sorts of reveries, and in the nineteenth century this made it a major tourist destination.

The published account of Banks's journal is mostly a physical description of the island. He was primarily a botanist and had no particular interest in minerals. He struggled at times to describe the rocks he found, at one point simply writing, "the rock is of a dark brown stone." But he collected a section of one of the hexagonal columns for later analysis and had his men take careful measurements of the cave—which he reported as being 117 feet, 6 inches high at the entrance (taller than a ten-story building) and extending 250 feet back into the island. Banks also measured the island itself, which is quite small, only about a mile long and a half-mile across.

Banks's detailed description of Staffa was an important contribution to knowledge, but he realized that that there was more to be learned: "The stone of which the pillars are formed, is a course kind of *Basaltes*, very much resembling the *Giant's causeway* in *Ireland*." The Giant's Causeway that Banks referred to is a formation of tens of thousands of interlocking basalt columns found along a three-mile stretch of the northeast Irish coast. It is one of the most recognizable geologic formations

Fingal's Cave in Staffa. This is the iconic image that Joseph Banks made for the account of Staffa he published in Thomas Pennant's *Tour in Scotland, and Voyage to the Hebrides 1772* (1774). It is only from this vantage point, in the waters directly before the cave's mouth, that one can see the full interior of the cave and marvel at nature's architecture. Smithson struggled to reach this vantage point, as did most nineteenth-century visitors. Courtesy of The Linda Hall Library.

in all of Europe, and engraved images of it were widely circulated in Smithson's time. The striking resemblance between the basalt columns in this formation and those found on Staffa was hard to miss.

The modern understanding is that the basalt columns at both locations were formed when a thick layer of molten lava flowed out of the earth and slowly cooled. And although the stone columns have a certain resemblance to crystals, the real reason they formed was that as the basalt slowly cooled and solidified, it also shrank and formed long vertical cracks. It takes a highly specific set of conditions to produce giant columns like those at Staffa, but this was the mechanism that formed them, and this is why the columns do not all have the same number of sides; as Banks noted, the columns at Staffa "are of three, four, five, six, and seven sides; but the numbers of five and six are by much the most prevalent."

In Smithson's time the question of how formations like those on Staffa were created was still an active topic of debate, and two fundamentally different processes had been suggested to explain them: the "neptunist" and "vulcanist" (or "plutonist") theories. Put simply, it was a choice between water and fire. The neptunist hypothesis held that the columns formed underwater, either through a process of sedimentation or, more likely, crystallization. Indeed, the organized shapes of the columns, so many of which were hexagonal, seemed to suggest that they were crystalline in nature.

But that scenario was challenged by the vulcanist explanation, which saw the columns as clearly the product of a volcanic flow that had somehow cracked into these shapes as it cooled, although the details of that process were still uncertain. What was not uncertain by the 1780s was the growing knowledge of the importance of volcanoes in the Earth's past, the growing evidence of some kind of "central fire" deep underground that fed them, and a growing acceptance of these ideas among natural philosophers.

Uno von Troil was an important member of the Swedish Academy and was with Banks when he first set foot on Staffa. He summarized the state of geological thinking in 1780: "Ten years ago it was a general opinion that the surface of the earth, together with the mountains upon it, had been produced by moisture. It is true, some declared the fire to be the first original cause, but the greater number paid little attention to this opinion. Now, on the contrary, that a subterraneous fire had been the principal agent gains ground daily: every thing is supposed to have been melted even to the granite." In less than two decades, this question of the "history of the Earth" would become one of the most divisive and hotly argued issues in all of science; but for now, it was still primarily a philosophical debate among gentlemen.

Smithson was likely familiar with this debate from his mineralogical studies and conversations with William Thomson, who had a strong interest in geology. It was also briefly discussed in Wall's chemistry class. Faujas was more intimately familiar with it, since he was one of its main participants. Early in his career he had developed an intense interest in volcanoes and became an advocate of the volcanic origin of basalt. His position at the Muséum National d'Histoire Naturelle had come through the support of the famous French naturalist Comte de Buffon, whose writings had made volcanoes a topic of special interest in French science. Additionally, Faujas had just published a scholarly book on the mineralogy of volcanoes and, in the book he planned to write after visiting Staffa, he hoped to demonstrate that Scotland was largely volcanic.

Smithson and his companions left London on August 29, 1784, at the unlikely time of six in the evening. Fortunately, the late start does not seem to have been a problem, since they were traveling by "post chaise." The chaise was the eighteenth-century equivalent of a taxi: a fast, lightweight, horse-drawn coach designed to carry two to four passengers. They were able to cover great distances because as the horses became tired they were exchanged for fresh ones at posts along the way.

Faujas hired three carriages: two for the four gentlemen and one for the four servants. After leaving London they traveled for ten hours straight, finally stopping at an inn at four in the morning only to get back on the road five hours later. They traveled long hours, often at night—a testament to the excellent roads between London and Edinburgh.

The group spent five days in Newcastle. Thornton had arranged visits to coal mines and other "manufactures," which Faujas was keen to describe in his book. They went down into two coal mines, carefully recording the different layers of rock as they were lowered into a shaft more than a hundred feet deep. This may have been Smithson's first experience in a mine, and he rode, as the miners did, standing in the same buckets that were used to bring up the coal. The buckets, in turn, were raised and lowered by a machine powered by four horses. This was how naturalists did fieldwork in Smithson's time, and he was getting invaluable training from one of the premier French specialists.

The next stop was Edinburgh, where there were prominent people to meet and much to explore. But it was already early September when they arrived, and they needed to keep moving before the weather changed. They stayed only long enough to hire wagons for the rest of their journey and make some brief introductions, though they all planned to return for a longer visit on the way back.

Next came Glasgow, where they visited more coal mines and explored the volcanic hills around the city. Faujas wrote of finding "some crystals of garnet, of twenty-four trapezoidal faces, of a greenish-grey colour, and very much resembling those found on Vesuvius. This is the first time that I have seen this kind of garnet in any other lava than those of Vesuvius. . . . I found near Glasgow only two specimens of the lave [lava] containing these garnets; M. de Mecies [Smithson] picked up a third."

They also visited John Anderson, professor of natural philosophy at the University of Glasgow, for whom they carried letters of introduction. In addition to his other projects, Anderson had established a natural philosophy course for tradesmen and written one of the first physics textbooks. Anderson's interest in workers was unusual for the time, and some years later, in 1796, he did something that may have engaged the interest of James Smithson. In his will Anderson left his entire estate "to the public, for the good of mankind, and the improvement of science, in an institution to be denominated Anderson's University." Anderson's request was honored, although the name was changed to "Anderson's Institution"— which was often called "The Andersonian." The similarity in both purpose and name between this institution and the institution Smithson later founded in the United States is noteworthy.

Faujas's party left Glasgow on September 14. They needed to reach Staffa before winter made passage to the island too dangerous to attempt. From this point on, the roads became much rougher, and finding accommodations for a group of their size was increasingly difficult. Faujas, unable to speak English and without a full appreciation of their situation, was beginning to lose control of the expedition; he must have been relieved when, after two days of hard and nearly continuous travel, they finally reached their next destination, Inveraray Castle.

The castle sits along the western shore of Loch Fyne, at the base of a low mountain. Despite its towers and faux moat, it was built to be a comfortable home for the Duke of Argyll and his family, and it was just being finished when the travelers arrived. The duke and duchess were famous for their hospitality and refinement. French was spoken at their dinners, food was prepared by an excellent French cook, and French wines, tableware, and table manners were employed at all times. Considering the difficulties the travelers had overcome to get there, coming across this kind of cultural island in such a remote location must have seemed like something out of a dream. But Inveraray was difficult to reach only by land. Despite its being narrow, Loch Fyne was deep enough for large oceangoing ships to sail straight up it and unload their cargos within sight of the castle. Everything necessary for the family's lifestyle came by ship, and much of it directly from France: art, furniture, visitors, artisans, food, and fine wines. The saltwater Loch Fyne also provided them with a rich selection of shellfish and seafood.

Like most educated people of the time, Smithson and his companions spoke French and were familiar with French customs. But they could linger only two days before resuming their journey. While still at Inveraray, Smithson wrote to a friend that his companions had encouraged him to drop out of the expedition and stay behind with the duke. Their rationale was that he was "delicate," which he immediately rejected, but the story suggests that all was not well within the group.

Before they left, the duke provided them with valuable advice about the best way to reach their goal. He recommended that they launch from the island of Mull, which was the closest point to Staffa and could be reached from the port of Oban. He gave them a letter of introduction to a man who lived on the west side of the island, in a house where one could actually see Staffa from the windows of the upper floors.

Following the duke's advice, the expedition headed north and then west to the small fishing village of Oban, but with just twelve miles left to go, the trip nearly ended in disaster. Faujas allowed the group to get caught on the road, at night, in a torrential storm. Unable to see, but trying to push forward, they lost their way and ended up going down a riverbed from which they could not retreat. With no other choice than to keep going, one of the wagons overturned, and the entire group ended up trapped in a muddy ravine. Fortunately, the townspeople heard their calls for help and rescued them, but the episode exposed Faujas's incompetence as a leader. By the time they arrived at Oban, the group was in disarray.

The next day Smithson and the others left Faujas behind and made their way to the island of Mull. There they were well received, but they had to wait several days until the seas were calm enough to allow them to attempt to reach Staffa. Finally, on September 24, 1784, the seas were quiet enough to permit passage. Smithson, his servant, his guide, and two sailors went in one small boat. Andreani, Thornton, their guide, and their sailors went in another.

Only a few entries from Smithson's expedition journal survive, and almost all of them relate to the events of the next few days. In the first entry he sets off from Mull: "Set off about half-past eleven o'clock in the morning, on Friday, the 24th of September, for Staffa. Some wind, the sea a little rough,—wind increased, sea ran very high,—rowed round some part of the island, but found it impossible to go before Fingal's cave,—was obliged to return,—landed on Staffa with difficulty,— sailors press to go off again immediately,—am unwilling to depart without having thoroughly examined the island. Resolve to stay all night." Despite the rough seas and the inherent danger, Smithson had insisted on trying to sail in front of the mouth of Fingal's Cave. This was the point from which Banks's famous illustration had been drawn, and this was the iconic view that subsequent visitors all wanted to experience, but Smithson was not able to reach it.

Both of the boats that carried the explorers to Staffa left as soon as they delivered their passengers. A storm was coming up, and Staffa offered little shelter for boats in bad weather, so they prudently withdrew to the island of Iona, some fifteen miles away, to wait it out. Smithson and the others stayed with one of two families living on the island at that time, sleeping on the straw in their barn.

The next morning was still stormy: "25th. Got up early, sea ran very high, wind extremely strong—no boat could put off. Breakfasted on boiled potatoes and milk; dined upon the same; only got a few very bad fish; supped on potatoes and milk;— lay in the barn, firmly expecting to stay there for a week, without even bread." In the afternoon Smithson was able to explore the island and gather some mineral samples and zeolites, but by the end of the day he must have realized that his dream of finding spectacular crystals like those Thomson had shown him was not likely to be realized. The best specimens had already been taken.

Luckily, on the third day the weather cleared and he was able to leave: "Sunday, the 26th. The man of the island came at five or six o'clock in the morning, to tell us that the wind was dropped, and that it was a good day. Set off in the small boat, which took water so fast that my servant was obliged to bail constantly—the sail, an old plaid—the ropes, old garters." They were all back on Mull by early afternoon, tired, hungry, and infested with lice, but otherwise uninjured.

Smithson seems to have rested only briefly after coming back from Staffa. Within a few days he and his servant were back in Oban, preparing to return to Edinburgh on their own. Smithson did not mention collecting mineral samples on Staffa in his journal, but he obviously did, because his first order of business after getting a room was to clean his "fossils" and pack them into a barrel for shipment to Edinburgh. (In the eighteenth century, the term "fossil" could refer to almost any unusual item brought up out of the earth, but in this context it could only refer to crystals.) He noted in his journal that the landlord charged him extra for the room "because I had brought 'stones and dirt,' as he said, into it." Smithson's samples, though not the display specimens he clearly hoped to find, were perfectly good for testing and chemical analysis, and nearly three decades later he would make good use of them in an article.

Smithson had already left by the time Faujas finally reached Staffa, and the Frenchman's journal confirmed that most of the island's crystals were already taken. The humid, protected environment of Fingal's cave provided the perfect place for them to grow, and over the course of many thousands of years a wide variety had formed in the spaces between the columns. However, Faujas reported that finding crystals was now "very rare in this cave." He was particularly keen to find "cubical" zeolites, and in his zeal he scoured the cave to gather any that remained, finally reporting that "having myself broken off all the specimens that I was able to see, I doubt whether those who may visit the place after me will find any quantity of it." Like all naturalists of the period, Faujas justified this kind of aggressive collecting as necessary for the advancement of science. Unfortunately, all the samples Faujas collected on his trip through Scotland, including the crystals from Staffa, were lost when the vessel on which he shipped them back to France sank off the Scottish coast.

The story of Staffa's crystals offers a good example of how ordinary science was conducted in Smithson's time, in that very little of what went on appeared in print. Joseph Banks did not mention the crystals in his article about the island, although he could hardly have missed them, and John Walker, William Thomson's professor at Edinburgh, never published an account of his visit to Staffa, although he showed the crystals to his students and talked about Staffa in his class.

Charles Greville, an avid London mineral collector, knew all about them; in a 1784 letter he referred to the "cubic Zeolithe, for which Stafva [Staffa] is most famed after the Collumns." Greville was keen to add Staffa's crystals to his collection, and he may have inadvertently been responsible for their disappearance. He had tried to visit Staffa a few months before Smithson but failed to reach it because of rough seas. In lieu of that, he sent "a man" to the island to gather specimens for him. That man may have been John Jeans, a Scottish mineral dealer, or "fossilist," as such merchants were then called. Jeans was well known to London collectors and for many years made an annual trip to London, bringing "exceeding fine" examples of "Scotch fossils" (minerals and crystals), which he sold at "great prices." He would have known how to select and handle the kinds of specimens that Greville desired, but as a mineral dealer he would have also been motivated to harvest as many of the valuable crystals as he possibly could. The theft of the crystals (for Staffa was privately owned at this time) was probably inevitable, given their value and vulnerability, but it happened with surprising speed and almost no evidence in print that they had ever existed.

Staffa's promise for shedding light on the geological questions of the day was never realized. Faujas was the first important mineralogist to visit the island, and he correctly deduced that it was the result of volcanic activity. But the publication of his book was repeatedly delayed and did not appear in English until 1799. By that time the neptunist-vulcanist debate had moved on to other questions, and his declaration that Staffa was an "extinguished volcano" elicited little scientific interest.

The true significance of Staffa seems to lie not so much in the realm of science, but rather in its reign as an icon of nature. "Nature's Cathedral," as it was frequently called, became a nineteenth-century destination—a place to encounter and contemplate the power and mystery of the natural world. Throughout the century, the remote little island attracted a steady and distinguished stream of visitors. The composer Felix Mendelssohn visited Staffa in 1829 and was so moved that he wrote *The Hebrides Overture* (also known as *Fingal's Cave*) to commemorate the experience. In 1847 Queen Victoria visited and had a contingent from the Royal Navy row her inside the cave so she could have the full experience.

For those unable to visit Staffa in person, recreations of it appeared in print and with surprising regularity in London's popular culture. In the 1790s two basalt columns from the Giant's Causeway in Ireland were prominently displayed on either side of the lobby in London's Leverian Museum (also called the "Holophusicon"), along with a long label explaining the details of the neptunist-vulcanist debate. One of these columns later appeared in a recreation of Fingal's Cave at William Bullock's private museum, the "Egyptian Hall." Representations inspired by Staffa also appeared on the London stage. There were at least two at the Sadler's Wells Theater, one in an 1805 pantomime and another, with water effects, in the 1810 drama *The Spectre Knight*. A representation of Staffa can still be seen today in the striking entrance to the Natural History Museum in London, which was directly inspired by the dramatic opening of Fingal's Cave.

For Smithson personally, although the expedition to Staffa failed to produce any new scientific discoveries, it gave him invaluable field experience, and the letters he and his companions sent from Scotland brought him to the attention of the London scientific community. The experience also boosted his confidence, and although he was only nineteen when it ended and still had to finish his education, the expedition to Staffa can be seen as the launching point of Smithson's scientific career.

———————+———————

2.

EDINBURGH, LONDON, AND PARIS

SMITHSON LEFT HIS COMPANIONS on Mull and does not seem to have had further interactions with any of them. He explained that he needed to return to London on urgent business, but once away from the group, he went straight to Edinburgh, where he stayed for nearly three weeks. He may have been hurrying to see people at the University of Edinburgh before they got busy with classes, and there were many people that he might have wanted to meet. This was the time of the Scottish Enlightenment, and Edinburgh was filled with famous and important figures, but Smithson spent almost all of his time with just two of them: Joseph Black, the famous chemist, and James Hutton, the soon-to-be-famous geologist.

Joseph Black was about as distinguished a figure as natural philosophy could produce in the late eighteenth century. In the first part of his career his long list of accomplishments included the discovery of "fixed air" (carbon dioxide) and the theory of latent heat—a fundamental scientific advance that played an important role in the development of the steam engine. Black was also interested in scientific instruments, and early in his career he made an important contribution to chemistry by inventing the "analytical balance," which brought a new level of precision to chemical analysis. Then, at the age of thirty-eight, he was named professor of medicine and chemistry at the University of Edinburgh, and the course of his career changed. From this point, although his chemical researches continued, he essentially stopped publishing and instead concentrated almost exclusively on teaching.

Black's move into the classroom would seem to have been a loss for science, but he was such an outstanding lecturer that, even from the classroom, he was able to exert great influence. Through his students and his many visitors, Black maintained an extensive network of correspondents with whom he exchanged information and materials and learned about the latest discoveries and theories—all of which he incorporated into his course. Indeed, he packed so much information into each lecture that it was not unknown for students to take the course a second time just to get it all. People came from all over Europe and North America to enroll in Black's course, and it is estimated that he taught chemistry to nearly five thousand students while in Edinburgh. Given that at one point the chemistry chairs at the universities of Edinburgh, Glasgow, Cambridge, and Oxford were all occupied by his former students, it is not hard to see how influential he was.

Black was fifty-six and well into the teaching part of his career when he met Smithson, who was nineteen. Smithson would have heard stories about Black and studied his discoveries as part of the Oxford chemistry course. He carried a letter of introduction from William Thomson that described Smithson in glowing terms, and the fact that it still survives among Black's papers is proof that they met. What they talked about during Smithson's visit is lost, but when Smithson got back to London he began a correspondence with Black that would continue for many years.

The other person that Smithson saw in Edinburgh was James Hutton, a good friend of both Black and Thomson. In another six months Hutton would give the first public reading of his revolutionary *Theory of the Earth*, which was nothing less than a completely new geological system. It was a project he had been working on for more than twenty years. If Smithson spent much time with Hutton, and it seems that he did, one of the things he would have seen was the mineral collection at Hutton's house. The collection, said to be perhaps the best in Edinburgh, was unusual in that the specimens had been collected primarily to support Hutton's geological theories rather than for their beauty or perfection. This point would have been impressed on Hutton's visitors by the large collection of boulders he kept in his yard. He acquired them on his extensive travels and used them as examples of how different layers of rock can distort and interact. This new and dynamic way to study geology was not easy. The boulders weighed as much as six hundred pounds each and had been brought by horse and wagon from sites hundreds of miles away.

If Smithson did visit Hutton's house, he could hardly have missed Salisbury Crags, an ancient lava flow that dominated the view from the backyard. In the 1780s the hard, basaltic rock at the top of the formation was being quarried for paving stones, and as the stone was removed, it constantly exposed new areas of the formation. Hutton hiked up to the site daily, where he took meteorological measurements and searched for clues as to how the different layers had formed. The rocks on Salisbury Crags showed that at some far distant time a layer of molten basalt, under great pressure, had forced its way under the existing layer of sedimentary stone and up into places where it was cracked. It was a stunning observation and implied the existence of underground lava, the welling-up of an immense ancient volcano, and then the passage of enough time for that volcano to have died out and eroded away.

This was the kind of geology that Hutton would soon propose: an endless cycle of the uplift of land, followed by weathering and erosion, and then uplift again—with no discernible beginning or end. He did not propose a mechanism for this uplift, although volcanoes seemed to be likely suspects, and he did not speculate on where the lava was coming from, although he found abundant evidence of its existence in the rocks of Scotland's mountains.

Hutton had a difficult writing style that undoubtedly slowed the adoption of his ideas. But in person, and in special locations like Salisbury Crags, another side of him came out, and from the evidence the rocks provided he could summon a compelling narrative of the land being transformed over unimaginably long periods of time. One of his colleagues, John Playfair, wrote of one such occasion when, as Hutton spoke, "The mind seemed to grow giddy by looking so far back into the abyss of time; and whilst we listened with earnestness and admiration to the philosopher who was now unfolding to us the order and series of these wonderful events, we became sensible how much further reason may sometimes go than imagination may venture to follow." Whether Smithson actually visited Salisbury Crags, meeting Hutton appears to have had a powerful effect on him. His lifelong interest in geology started around this time, and when he began to write his own articles, his arguments and observations consistently supported positions Hutton would have favored.

Before Smithson left, both Black and Hutton made requests of him. Black asked him to find a sample of amianthus, a satiny form of asbestos, and Hutton gave him two guineas to procure and send him fossilized bones from the quarries around Oxford. Neither of these requests was unusual; Hutton and Black made similar requests of their other correspondents and would in turn try to respond to any requests they received. For his part, Smithson seems to have been delighted with the arrangement. He stayed in touch with both men for the rest of their lives, and their surviving correspondence shows mutual affection and respect.

After leaving Edinburgh, Smithson was in no hurry to return to London. It was getting toward the end of October 1784, and he and his servant now traveled on horseback, riding up to Leadhills and the neighboring village of Wanlockhead, about forty-five miles southwest. It was not the ideal time of year to visit this high, windswept area, but Smithson had a special purpose for going to these villages. They sit atop a massive deposit of lead ore, and there are also significant deposits of gold, copper, and zinc in the area, all of which were being mined in the eighteenth century. As the miners dug deeper and further, following the veins of ore back into the hills, they sometimes came across unusual new minerals, many of which were then found nowhere else in the world.

Smithson's goal was to obtain a specimen of a curious and very rare mineral that Black had told him about and which he may have seen in Black's mineral collection. It was a noticeably heavy, cream-colored material, sometimes forming crystals, sometimes not, that Black called *Terra Ponderosa Aërata*. Black did not have enough to share, but he suggested that Smithson see his contact, the manager of one of the Leadhills mining companies, who saved unusual specimens for Black. Smithson wanted this mineral for his collection and experiments, but rare natural materials like this could also gain one introductions and other benefits in London's scientific world.

Known today as witherite (barium carbonate), it was a mineral of interest for several reasons. The first was its weight. With a specific gravity of more than 4.3, it was no wonder that it was called *Terra Ponderosa*, Latin for "heavy earth." Even a small piece of it felt unusually heavy; by comparison, granite has a specific gravity of less than 2.7 and slate around 2.8. Its second interesting characteristic was the *Aërata*, the surprisingly strong effervescence that started after placing just a few drops of acid on it. The light-colored mineral contained a surprisingly large amount of "fixed air" (carbon dioxide), a gas that could be dramatically released by dropping an entire piece of the mineral in hydrochloric acid. Interestingly, dissolving it in sulfuric acid converted it to a more common mineral, which Black called *Gypsum Ponderosum* and is now known as barite (barium sulfate). This was noteworthy, because in eighteenth-century chemistry metallic minerals could often be changed in this way, and this was what made *Terra Ponderosa Aërata* particularly interesting—it was widely suspected of containing a new metal.

William Withering had already submitted a paper to the Royal Society hypothesizing this. Although the discovery of a new metal was widely expected, neither he nor anyone else had been able to isolate it—to refine it to the point where it could be seen by itself, uncombined with anything else. We now know that this was an impossible task given the chemical technology of the late eighteenth century. Barium is so reactive that it cannot exist by itself under normal conditions on the surface of the earth; it can exist only in combination with something else. It would be another twenty-four years before Humphry Davy finally isolated it—but only briefly, and only with the aid of a powerful chemical battery. Davy ran a large electrical current through a mixture of barium oxide and mercury, which softened it and broke the barium-oxygen chemical bond. Oxygen was then drawn to one terminal, where it bubbled away, and pure barium was drawn to the other terminal, where it accumulated as a shining, silvery metal that quickly combined with oxygen as soon as the power was disconnected.

But in 1784, all this was still far in the future, and it was a challenge to get a sample of *Terra Ponderosa Aërata* to experiment on. Mineral dealers did not carry it, and Smithson had not secured any in Leadhills. It was a discouraging trip for him, and by the time he finished there he had been traveling continually for more than two months. Yet, instead of returning to the familiar comforts of either Oxford or London, he made a detour to the Northwich area, southwest of Manchester, to visit a salt mine.

Smithson was developing a taste for travel, and he did not reappear in London until well into November. The letters he had sent to his friends from Oban, after visiting Staffa, had arrived a few weeks earlier and were being circulated, which may have given him a brief celebrity. Of much greater interest in scientific London was the report he brought back from Edinburgh, particularly the news about *Terra Ponderosa Aërata*, information that he would use a few months later at his first public appearance.

His energetic friend William Thomson had arranged for that appearance. While Smithson was still on the road to Staffa, Thomson nominated him for membership in the Society for Promoting Natural History. This was one of a number of London science clubs that were active in the city's coffeehouses, this one meeting monthly at the Black Bear in Piccadilly. Thanks to Thomson's sponsorship, Smithson's membership was waiting for him by the time he returned. All that remained was for him to attend a meeting to be interviewed, followed by a vote on his membership. In early 1785, Thomson came down from Oxford to accompany Smithson to the meeting.

Smithson brought a sample of *Terra Ponderosa Aërata*, which he had somehow procured after his return to London. It made a big impression. The ensuing discussion gave Smithson a chance to demonstrate how much he knew about it, and by the end of the meeting he was not only a member but also elected to the society's mineralogy committee. His debut into the London scientific community was an unqualified success.

Smithson was strongly drawn to the scientific life that was opening for him in London, but first he needed to finish his degree at Oxford. He was now in his third year and would spend most of it at the university, but his fourth year was more flexible, and for most of it he shuttled back and forth between the two locations, advancing his scientific career while completing his studies. Take, for example, his schedule in the spring of 1786. In March he attended a Royal Society Club dinner. These dinners were often a first step in being invited to join the society, and this opportunity enabled him to meet a distinguished group of Royal Society members. The following month, at the invitation of noted chemist Richard Kirwan, Smithson attended his first meeting of the Coffee House Philosophical Society, a select group that Kirwan had established and presided over.

On May 26, 1786, Smithson graduated from Oxford. He could have continued on in school, although the only advanced degrees in chemistry offered in Smithson's time were actually medical degrees specializing on a chemical topic. But Smithson showed little interest in medicine and none in remaining in school. He was anxious to move on to a life in London. Once settled, he quickly became a regular at Royal Society meetings, usually attending as the guest of Kirwan. Smithson soon had a number of other key supporters in the Royal Society, including the president, Joseph Banks. Banks was famously social, and in addition to seeing him at meetings, Smithson probably attended events at his house. Smithson's first apartment in London was just a five-minute walk from Banks's residence.

Banks would later assist Smithson in important ways, but his real mentor during the early days in London was Henry Cavendish. The two formed an unexpected friendship, and over the course of the next few years Cavendish would have a significant influence on Smithson's scientific life. He was fifty-five when Smithson met him and widely acknowledged to be a complex individual. Humphry Davy called him "a great Man with extraordinary singularities," and any account of him needs to convey that he was brilliant, extremely wealthy, and probably the best experimental philosopher in England, but also that he had a number of unusual personal characteristics that could make social interaction with him difficult.

Cavendish was extremely taciturn. Indeed, one biographer described him as having "uttered fewer words in the course of his life than any man who ever lived to fourscore years, not at all excepting the monks of La Trappe." He seems to have suffered from extreme shyness and as a result found it difficult to meet new people. The chemist and physician Thomas Thomson, also a Royal Society member, described him as "shy and bashful to a degree bordering on disease; he could not bear to have any person introduced to him, or to be pointed out in any way as a remarkable man." Thomson told the story of Cavendish being introduced to a foreign visitor at an event in Banks's house, and having to listen to a long, flattering speech praising his scientific accomplishments. As the man spoke, "Mr. Cavendish answered not a word, but stood with his eyes cast down, quite abashed and confounded. At last, spying an opening in the crowd, he darted through it with all the speed of which he was master, nor did he stop till he reached his carriage, which drove him directly home." Cavendish was even more uncomfortable around women, avoided contact with them, and never married.

To his credit, he did not let any of this keep him from pursuing an active life of science and public service, even though the requisite social interactions were never easy for him. His colleagues tried to help by instructing visiting scholars on the best ways to approach him. Banks once advised a guest interested in Cavendish "to avoid speaking to him as he would be offended," but that if Cavendish spoke first then he should continue the conversation, because Cavendish was "full of information." At social events he would sometimes move quietly from room to room, occasionally pausing to listen in on the conversation of a particular group and contributing comments. For situations such as these, Smithson's friend William Wollaston advised that "the way to talk to Cavendish is never to look at him, but to talk as it were into vacancy, and then it is not unlikely but you may set him going."

Smithson first met Cavendish in the spring of 1786 at a dinner of the Royal Society Club, a select group of Royal Society members who met before each Society meeting. It was common for members to bring interesting guests, and Faujas had been invited to one of these dinners before he left for Scotland. Cavendish appears to have invited Smithson to the dinner to hear how the expedition turned out, and, despite the famous man's shyness, the two soon became friends. As his assistant later recalled, "Mr. Cavendish liked him very much, & seemed to take great pleasure in his company."

Among Cavendish's many accomplishments was the discovery of hydrogen—which Smithson had studied in Oxford. Cavendish had already received the Royal Society's prestigious Copley Medal for his pioneering work on gases, and in 1785, he published the results of experiments that showed nitrogen to be a component of atmospheric air. It was an important discovery, and when other chemists had trouble replicating his results, he felt obliged to publicly repeat the lengthy experiment and have it monitored by a select group of witnesses, among them Banks and Smithson.

Cavendish had also prominently supported Smithson a few months earlier when he was proposed for membership in the Royal Society, and on April 18, 1787, less than a year after he graduated from Oxford, Smithson was unanimously elected a Fellow. This was the highest level of English science—and he was just twenty-two years old.

Today, candidates for fellowship in the Royal Society are expected to have made "a substantial contribution to the improvement of natural knowledge," but this was not a requirement when Smithson was a candidate. Science at that time, especially in England, was still mostly conducted in person, and Royal Society members often voted as much on their personal knowledge of the candidates as on their resumes. So a young man with unusual potential, one who was impressive in person but had not yet accomplished very much—like Smithson—could have legitimately been elected a Fellow.

This is what we see in the description of Smithson on the certificate proposing him, which makes no mention of any accomplishments: "James Lewis Macie Esq. [Smithson], M.A. late of Pembroke College Oxford, & now of John Street Golden Square, a Gentleman well versed in various branches of Natural Philosophy & particularly in Chymistry & Mineralogy, being desirous of becoming a member of this Society, we whose names are hereto subscribed do from our personal knowledge of his merit, judge him highly worthy of that honour & likely to become a very useful & valuable member."

No detailed description of Smithson's personality has survived, but this image of him as an engaging, earnest young man of great potential invites further consideration. The only personal account of him comes from Cavendish's assistant, Charles Blagden, who once described him as "an excellent Chemist, with much information in many other parts of Science, & great precision of ideas" and recalled that "I was frequently with him, and scarcely ever without acquiring some new ideas."

By "ideas," Blagden was almost certainly referring to Smithson's deep interest in chemical theory, and this was what really seems to have set him apart. Chemistry had two sides in Smithson's time, practical and theoretical. While Smithson was an outstanding practical chemist, his true interest was always in the theoretical side. His abilities were such that in early 1788 Joseph Banks could write that Smithson "far exceeds his peers in his knowledge of philosophical chemistry." Banks wrote this in a letter of introduction for Smithson to give to Antoine Lavoisier, the great French chemist, whom Smithson had long been keen to visit. He had to wait until Cavendish's nitrogen experiment was finished, but on March 19, 1788, the witnesses had their last meeting. They were able to confirm all of Cavendish's original findings, and shortly thereafter Smithson left for France.

He arrived in Paris in July 1788, traveling with a group of other young men with similar interests, several of them also Royal Society members. Smithson's desire to meet Lavoisier was understandable. Brilliant and well-connected, Lavoisier was at the center of a French movement that was revolutionizing the science of chemistry, and Smithson had long studied his work. Just a few months earlier, as a gesture to acknowledge his achievements, Lavoisier had been elected a foreign member of the Royal Society, and Smithson seized on the idea of offering his congratulations in person as an excuse to visit him. He carried a letter of introduction from Joseph Banks, and he had just witnessed Cavendish's repetition of an important experiment on gases. He had every reason to think that Lavoisier would be interested in him.

No record of the visit seems to have survived, but they likely met, because Banks's letter of introduction is among Lavoisier's papers. However, there is no mention of Smithson in any of Lavoisier's other papers, and no evidence that the two men corresponded afterward. Smithson does not seem to have made an impression on the great man, but if he failed with Lavoisier, the same could not be said of his encounter with the mineralogist René Just Haüy, still remembered today as "the father of crystallography."

In 1784, Haüy had proposed an elaborate theory of crystal formation that seemed to reveal some basic truths about the nature of matter. The theory was very technical, based on the measurement and mathematical analysis of the face angles of crystals. It assumed the existence of certain "integrant molecules," submicroscopic forms with specific geometric shapes that, under certain conditions, could physically fit together to make the different types of crystals found in nature. Today it may best be seen as a kind of interim theory of atoms, but in 1788, when Smithson met him, Haüy's theory was at the cutting edge of science, and Smithson embraced it enthusiastically. After his return to London, the two men began a vigorous correspondence that would continue for more than three decades. Smithson met other important French savants during his short stay in Paris, including the influential chemist Claude Louis Berthollet and the pioneering naturalist Georges Cuvier, with whom he would later correspond, but it was Haüy who had the biggest influence on him.

Smithson left Paris toward the end of September 1788 and returned to London. It was a good time to leave, since France was already starting to show signs of what would soon explode into the French Revolution. Smithson returned to the familiar activities that had occupied him before he left, such as building his mineral collection, keeping up with his scientific reading, reproducing interesting experiments, and maintaining his activities with the Royal Society. But there were some changes as well. Smithson completely reorganized his mineral collection when he got back, abandoning the large showy specimens that characterized most English collections and began collecting small, perfect crystals, as Haüy advocated. The change was so dramatic that he no longer had any use for the spacious storage cabinets he had commissioned before he left, and he sold them to another collector without ever having used them.

Smithson was making studies of different mineral groups, such as the olivines and the ores of lead. He made detailed comparisons of individual samples and, for the first time, he measured the angles of their crystals, just like Haüy. None of these mineral studies survive, and Smithson does not seem to have published any of them, but he apparently sent copies to Haüy, because references to them began to appear in Haüy's works. Of particular importance was Smithson's independent discovery of an unusual crystal form of the mineral calcite and his subsequent demonstration that it could be explained by Haüy's crystallography. Haüy promoted this research as an important confirmation of his theory, and it brought Smithson's name into broader scientific notice.

Smithson was also interested in the electrical properties of crystals, a topic that Haüy himself was investigating. Smithson is reported to have undertaken an extensive study of these crystals, which he failed to publish, but this interest in electricity would reappear in several of his later articles. Crystallography was Smithson's new passion, but he continued to be interested in other topics. In 1790 a new substance came to his attention that—for a while, at least—engaged him completely.

———————————————

3.

TABASHEER

AT THE ROYAL SOCIETY'S MEETING on March 11, 1790, a letter was read that profoundly affected James Smithson's scientific career. The letter was from one of the society's members, Patrick Russell, who sent it from Vizagapatam (now Visakhapatnam), a city on the east coast of India. Russell worked there as a physician for the East India Company, and he described a curious form of materia medica made from small stones found inside bamboo plants. The stones were called "tabasheer," and Russell reported that they were found inside the chambers of the plants—where they could have been made only by the plant itself. Along with his letter, Russell sent a package of tabasheer samples and seven bamboo "reeds" that he suspected of containing more.

Seven small bags of tabasheer, each collected from a different place or at a different time, awaited inspection. Russell had collected the contents of one directly from a bamboo plant, and a man who worked with bamboo had brought him another. Russell bought three more bags in a market "at a very low price," while one of the bags was not identified. Russell purchased the largest batch at Hydrabad (now Hyderabad), a city in southern India, and reported them to be "the finest kind of this substance to be bought." Four of the bamboo reeds were also carefully split open at the meeting, and the members judged their contents to be the same material as the samples in the bags.

As a mineral produced by a plant, tabasheer did not fit easily into any of the familiar categories of matter: animal, vegetable, or mineral. This system, which classified materials on the basis of where they originated, was introduced into chemistry in the seventeenth century, and it carried within it the assumption that these were distinct forms of matter. By Smithson's time these categories had been largely condensed into what we now call "organic" and "inorganic" materials, but it was still widely believed that substances in these categories were produced according to fundamentally different chemical "principles." Indeed, it was a common belief that many organic compounds could only be produced inside living bodies. As the English chemist William Lewis wrote in 1773, "No art can prepare or extract from the substances by which vegetables are supported, products in any respect similar to those elaborated in the bodies of vegetables themselves."

The inability of Enlightenment chemists to produce organic compounds from inorganic materials seemed to support the idea of a special force within living bodies, and the question of whether such a "vital" force actually existed was a legitimate scientific question throughout most of Smithson's career. Because tabasheer was a mineral of "vegetable" origin, it seemed perfectly positioned to shed light on the organic/inorganic distinction: if it turned out to be a known mineral, then the distinction would seem not to be absolute, but if tabasheer turned out to be a new material, it would support not only the distinction but also, perhaps, the existence of some kind of vital force. The analysis of these strange stones therefore needed to be as accurate, complete, and objective as possible, and Joseph Banks chose James Smithson to perform it.

Smithson was only twenty-five at the time and had yet to publish any articles, but his keen interest in the philosophical side of science may have been what made Banks select him. It was a plum assignment, and the only problem seemed to be whether or not he was available. Smithson had moved into a house on Orchard Street just a few months earlier, and on March 20, just nine days after the reading of Russell's letter, Banks's secretary (and Henry Cavendish's erstwhile assistant) Charles Blagden reported that Smithson was still "setting up his library and laboratory." This was a problem, but one that they apparently solved, because in mid-July it was announced in the Royal Society's journal that "several specimens [of tabasheer] are now under chemical trial."

The chemical analysis that Smithson embarked on was based largely on observations and the production of effects. Materials were identified by such things as color, hardness, or taste and by the way they reacted to high heat, water, or acids. There was little underlying theory in eighteenth-century chemistry, and the measurements chemists were able to make were almost exclusively confined to weight and the related measurement of specific gravity. Still, chemical analysis in Smithson's time had improved significantly from where it had been at midcentury, thanks mainly to the development of "wet analysis." This method involved the systematic application of liquid acids and liquid alkalis to break down materials and get them to dissolve. Once a substance was in the liquid state, chemists had a variety of techniques to isolate and identify its individual components.

By Smithson's time, wet analysis had developed to the point where numerous commonly used tests could reliably identify most of the known materials. These tests had been collected and organized into a number of practical manuals that Smithson used as the starting point for his investigation. Chemical analysis was still highly dependent on the skill of the chemist, but these books at least ensured a common methodology.

On July 7, 1791, more than a year after Russell's tabasheer was displayed to the members, Smithson's analysis of it was read to the Royal Society. Titled simply "An Account of Some Chemical Experiments on Tabasheer," it appeared in the society's journal a few months later. This was Smithson's first publication, and he had taken great pains to make it as comprehensive as possible. His elaborate analysis consisted of more than sixty separate tests. Even apart from the differences in terminology, understanding Smithson's article is not easy for the modern reader. Smithson was a knowledgeable investigator, writing for a relatively small and knowledgeable audience. He did not provide details about the tools he used, nor did he interpret the results of the tests he made with them. He described what he saw and expected his readers to know what it meant. Smithson's description of his analysis of tabasheer is more detailed and complete than any of his other published analyses, and it provides the modern reader with some insights into the way he accumulated information and ultimately used that information to reach a conclusion.

Smithson's tests on tabasheer can be grouped into seven different sections, which makes them a bit more accessible. The initial section was essentially a "first pass"—a series of mostly nondestructive tests whose results would be used to shape the subsequent inquiry. The second section tested the tabasheer with water, and the third tested it with "vegetable colours," indicators of acidity and alkalinity similar to modern litmus paper. Most of the remaining tests were destructive. Those in the fourth section tested the samples with heat, and those in sections five and six subjected them to "wet analysis." In the seventh and final section, the tabasheer was combined with different kinds of "fluxes" and heated to make "glass."

Smithson first started with a physical description of the tabasheer. He decided to work exclusively with the samples Russell had purchased in Hyderabad, which he considered to be of superior quality. The pieces themselves were irregularly shaped and small, the largest about the size of a small pea. Some of the pieces "bore impressions of the inner part of the bamboo against which they were formed," evidence that this was indeed a product of nature. Smithson also noted tabasheer's resemblance to "cacholong," a kind of opal, and that some of his samples were opaque and absolutely white, while others were "pellucid [transparent] and had a bluish cast." These observations were suggestive: "earths" tended to be white or light-colored, and opals were known to be composed mostly of "siliceous earth"—silica.

Smithson reported being unable to break the tabasheer by squeezing it between his fingers but noted that it was easily reduced to powder by grinding it between his teeth: "On first chewing it felt gritty, but soon ground to impalpable particles." He also noted that some of the particles adhered to his tongue and that the tabasheer "had a disagreeable earthy taste, something like that of magnesia," a chalky mineral sometimes used as a stomach medicine.

While chewing and tasting may seem odd and even quaint, this kind of description was traditional in eighteenth-century mineralogy. In the absence of photography and quantitative measurements, descriptions like this were the best way for mineralogists to accurately identify the materials they were writing about. In Smithson's time, it was widely acknowledged that an individual with an acute palate could often detect substances in small amounts that were beyond the reach of chemical analysis. Accordingly, lists of standardized descriptive terms for color, hardness, and taste, complete with instructions on how to use them, were often included in mineralogical texts.

Smithson's next tests were performed in the dark. In the first, he cut one of the samples in half with a knife. He was looking for a spark, which would have been an indication that the sample contained flint, a form of siliceous earth. "Striking fire with steel" had long been one of the identifying characteristics of minerals containing flint, but Smithson reported that "no light was produced." He also rubbed the two pieces together, but to no effect. These tests did not rule out the presence of silica in the tabasheer, but it was not present in the form of flint.

He then tested tabasheer for what, at this time, was called "phosphorescence." This test also took place in the dark and consisted of placing a piece on a hot iron. This gentle heating made the sample glow slightly, and Smithson described it as "surrounded with a feeble luminous *aureole*," but the effect was slight and disappeared entirely after the tabasheer was made red hot. This test had been suggested by Bergman as an indicator of "Calx fluorata," the mineral fluorite, which when "exposed to heat, below ignition . . . emits a phosphorescent light." This property of fluorite was so distinctive that the term "fluorescence" was named after it, but tabasheer did not have this property and thus did not contain fluorite.

Smithson carefully examined a specimen under a microscope, but he reported that "it did not appear different from what it does to the naked eye." His "microscope" may have been an early version of the familiar laboratory microscope, but it was more likely a "simple" microscope that used only one lens and had a relatively low magnification.

The last test in this series was a measurement of tabasheer's specific gravity. This is a measurement of a material's density compared to the density of water, and it was one of the few quantitative measurements that chemists of this period were able to make. It actually consisted of two measurements: the weight of the tabasheer when dry divided by its weight when submerged in water. It revealed that tabasheer had a very low density, lower than almost any other mineral. Smithson measured it at 2.188, which effectively ruled out a large number of substances, especially the metals, which by definition have a specific gravity above 4.9. It was also significantly lower than that of quartz, which was measured as 2.66 and thought to be the purest form of siliceous earth.

Because the specific gravity measurement was so low, and because his reputation as an analytical chemist was not yet established, Smithson took the unusual step of asking Henry Cavendish to repeat the measurement. When Cavendish's result of 2.169 turned out to be even a bit lower than Smithson's, it provided him with important support. Smithson's defense of his measurement also included a postscript to the article, with details about his methodology. Because tabasheer is porous, it might have seemed possible that air trapped inside it could have affected the measurement of its weight in water, but he assured his readers that "great care was taken in both the experiments that every bit was thoroughly penetrated with the water, and transparent to its very center, before its weight in the water was determined."

With the preliminary investigation out of the way, Smithson now subjected the tabasheer to tests with water. This was the starting point for most eighteenth-century chemical analysis, and these tests were all performed with distilled water, which Smithson probably made himself from rainwater. In the first two tests, Smithson simply describes putting pieces of tabasheer in water. They "emitted a number of bubbles of air; the white opaque bits became transparent in a small degree only, but the bluish ones nearly as much so as glass. In this state the different colour produced by reflected and by transmitted light was very sensible." The colors that Smithson observed, blue by light reflected off it and a flame color when light passed through it, would later be used by him as one of tabasheer's identifying characteristics.

In the second test, pieces of the bluish tabasheer were weighed and then put into water until they became transparent. On being "taken out, and the unabsorbed water hastily wiped from their surface," they were weighed a second time and found to have absorbed their weight in water. This property may account for tabasheer's reported use in the treatment of smallpox. Before the introduction of the smallpox vaccine, people who contracted this disease suffered devastating scars from the sores it produced. Because of its remarkable ability to absorb liquids, powdered tabasheer may have been effective in drying these sores and reducing the subsequent scarring. In any case, absorption was a characteristic of several of the earths, particularly siliceous earth.

In the next test, Smithson boiled four small pieces of tabasheer in distilled water. Working on a very small scale (the total weight of his sample being just 3.2 grains), he placed them in just one-sixteenth of a cup of water. After thirty minutes of boiling he combined samples of the water with both acids and alkalis, but observed no effect. After the water evaporated, however, he found a white film inside the flask that could only be removed by boiling it in a strong alkali—clear evidence that it was siliceous. Repeating this test with the same sample produced the same result, which was puzzling because silica had long been thought to be completely insoluble in water. But it had been shown to be quite soluble in *alkaline* water, with which it formed a solution known as "liquor silicum" and which would leave a white deposit when it evaporated.

In his last test with water, Smithson boiled a tiny piece of tabasheer (just three-tenths of a grain this time) in more than a quart of "soft water." His stated object was to see "whether the whole of a piece of Tabasheer could be dissolved," but his use of soft—that is, nonalkaline—water seems to indicate that he thought the water used in the previous test had been slightly alkaline. In any case, after boiling it for almost five hours he found that the tabasheer "was not diminished in quantity, nor was it deprived of its taste" and he did not report finding any white film.

These tests all indicated that tabasheer was a form of siliceous earth. But this kind of analysis depends on an accumulation of evidence, and there were many other things to be tried. The next series of tests used plant substances, or "vegetable colours," to determine the tabasheer's "saline contents" (now called pH). These plant substances changed color in very specific ways when put in contact with acids or alkalis. A great number of these indicators had been discovered, and because they varied in sensitivity it was common to use at least two of them in an analysis. Smithson chose three of the most common: logwood, turnsole, and dried red cabbage.

Logwood is a native tree of Mexico, whose valuable wood had been imported to Europe since at least the seventeenth century. It was primarily used as a dye and, depending on how it was processed, was capable of producing a variety of vibrant colors for both cloth and paper. It was also used as a medicine, and Smithson probably got his supply of it from a London apothecary. To make a medicine, the patient was instructed to steep pieces of the wood for several hours in warm water and then drink the resulting brownish liquid. Smithson prepared his logwood solution in the same way, and after dropping a bit of the tabasheer into it, boiled it for a long time. If the tabasheer was acidic it would have turned the liquid red, and if alkaline, blue, but the color of the solution did not change, which showed that it was neutral.

Turnsole was another natural dye, this one made from the fruit of *Chrozophora tinctoria*, a type of heliotrope plant. It too was capable of producing a variety of colors, depending on how it was treated. It was produced in large quantities in France, where the raw dye was extracted from the plants, processed, then soaked into strips of muslin and finally dried. Smithson soaked one of these "rags" in water to make a light blue solution that turned red in the presence of acids and yellow when exposed to alkalis. But after boiling the tabasheer in it "for a considerable time" the solution's color remained unaltered, again showing it was neutral.

Dried red cabbage was the final material that Smithson used, and he probably made it himself. Red cabbage was widely used for these kinds of tests because it was inexpensive, easy to use, and sensitive to a broad range of acids and alkalis. It was best when made from fresh red cabbage, which when chopped up and soaked briefly in warm water produces a blue liquid that turns red with acids and green with alkalis. When fresh cabbage was not available, dried strips of it could be used in the same way, although with some loss of color and sensitivity; the fact that Smithson used dried red cabbage probably indicates that this test was made at a time of the year when fresh cabbage was not available. As with the other two indicators, boiling the tabasheer with red cabbage produced no change in its color, the result one would expect if tabasheer was a form of siliceous earth.

Subjecting the samples to fire (and heat) were traditional ways for chemists to analyze unknown materials, and the traditional way to do this was with different kinds of furnaces, each designed to produce a particular high-temperature environment. Typically, the specimen would be placed in a ceramic crucible, covered, and then placed in the furnace. The crucible remained covered while the furnace was working and was usually only opened after it had cooled. It was a powerful method, especially when working with metals, but not being able to see or manipulate the sample during the test was a major disadvantage. By Smithson's time, the furnace had been replaced for many analytical purposes by two different tools, the forge and the blowpipe.

The first four fire-based tests were performed with a forge and in small "earthen" (porcelain) crucibles. The forge Smithson used would probably have been similar to that used by blacksmiths: an open metal tray filled with coal, mounted at a convenient working height, and attached to a bellows apparatus that blew air into the open fire and dramatically raised its temperature.

In the first test, an uncovered crucible was placed in the fire and made red hot. A piece of tabasheer was thrown into it and observed. It did not burn or even blacken. Indeed, Smithson reported that "it underwent no visible change." After it cooled it was harder and had lost its taste, but when he put it in water it still became transparent, an indication that it had not melted in any way. In the second test, he put several pieces in a crucible and heated them until they were red hot. After they cooled he weighed them and found that they had lost two-tenths of a grain. He attributed this loss to "the expulsion of interposed moisture" and reported that they regained the weight "on being exposed to the air for some days." In the third test, a piece of tabasheer was put into a crucible "surrounded by sand, and kept red hot for some time." When it cooled he cut it up and examined it, inside and out, but could detect no changes. In the final test, he melted niter (potassium nitrate) in a crucible until it was red hot and then threw in a piece of tabasheer. The use of niter was a common technique to see whether a substance could be induced to ignite, but Smithson reported that it did not burn "or seem to suffer any alteration." Commonly called saltpeter, niter (nitre, in the British English spelling) was one of the ingredients of gunpowder and had been studied in the seventeenth century by the noted chemist Robert Boyle. During his pioneering work with vacuums, Boyle discovered that gunpowder would burn—but not explode—in the absence of air, and he concluded that heating niter produced "agitated vapours which emulate air" and thus encouraged combustion. The fact that tabasheer would not ignite even under these extreme circumstances was convincing evidence that it was completely resistant to combustion.

Having finished his tests with the forge, Smithson now subjected bits of tabasheer to the heat of the "blowpipe." A blowpipe is essentially a short, tapered metal tube, bent at one end, that is used to blow into the flame of a candle or lamp. As oxygen in the user's breath combines with the flame, it becomes much hotter, and the blowpipe's shape allows the user to direct the flame onto the object being tested. In the hands of a skilled chemist such as Smithson, this simple instrument could produce a long blue flame that was surprisingly hot. It easily melted brass and, unlike the flames of the forge, could be targeted directly onto the sample—which Smithson did. To protect his hands from the heat, he held the tabasheer on a piece of charcoal. The resulting test was conducted less than a foot away from his face and gave him a clear view of what was happening. He reported that the tabasheer initially "diffused a pleasant smell" and that it contracted noticeably. It also briefly became transparent, but "seemed not to shew any inclination to melt *per se*." Smithson suggested that the blowpipe may simply have softened it and, as evidence, noted that existing cracks in some of the pieces had been widened by the intense heat but that none of them had fallen apart. This agreed well with the test for silica in Axel Cronstedt's *An Essay towards a System of Mineralogy* (1788), one of the books in Smithson's library: "After being burnt, it does not fall to a powder . . . but becomes only a little looser and more cracked by the fire." Smithson also noted that when some ash from the charcoal he used to hold the tabasheer fell on it, "it instantly melted and small very fluid bubbles were produced." In order to ensure that unseen alkaline ash would not affect his test, he repeated it. But this time he held the tabasheer on the end of a glass tube "in the method of M. de Saussure," and reported that it did not melt.

In his widely read *Outlines of Mineralogy*, the Swedish chemist Torbern Bergman listed the characteristics of "terra silicea" (silica), one of which was that "when pure, it is refractory [unaffected] in the fire." By the time he reached this point of the analysis, Smithson must have been fairly confident that his tabasheer was almost completely composed of silica. But the true test would be how it performed in wet analysis, and one of the most distinctive traits of silica was that it was not affected by any of the acids. The only exception to this rule being "fluor acid" (hydrofluoric acid), which was said to readily dissolve it. Accordingly, Smithson now tested his tabasheer by boiling it in acids. He first tried "pure white marine acid" (the strongest form of hydrochloric acid), but before putting the pieces of tabasheer into it, he first soaked them in distilled water. He explained that this was done "to expel the common air contained in them," which otherwise would have created bubbles and made it impossible to tell whether the acid caused the release of any gas. This precaution was undoubtedly in response to Cronstedt's observation that pure silex "excites no effervescence with acids." Smithson dutifully reported that "no effervescence arose on its immersion into the acid." Bringing the acid to a boil also seemed to have no effect. After letting the acid evaporate, he washed the sample in distilled water, heated it red hot to get it dry, and found that it had not been reduced in weight.

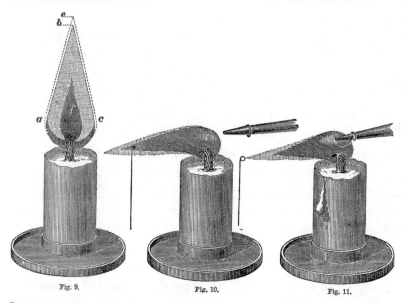

The blowpipe flame. To perform "blowpipe analysis," it is necessary to understand the different parts of a flame (left). Placing the tip of the blowpipe on the side of the flame and blowing gently produces a "reducing flame" (center), which is useful for the analysis of metals. Placing the tip of the blowpipe in the flame, just above the wick, and blowing harder produces a "blue flame" (right), which is much hotter. It was also important to know where in the flame to place the sample. Courtesy of the Smithsonian Libraries.

He repeated the test with a larger sample. This test was identical to the previous one, except this time he omitted putting the tabasheer in water, and after boiling it in acid and then washing it, he let it dry in the air for several days rather than heating it. It made no difference; there was still no perceptible loss of weight. The same test was also repeated with "pure white nitrous acid" (strong nitrous acid) and then with "strong white vitriolic acid" (concentrated sulfuric acid), but in both cases the tabasheer showed no change "either in its weight or properties." In an attempt to induce the tabasheer to react, Smithson used a special technique whereby he formed a paste by mixing a small amount of powdered tabasheer with a few drops of sulfuric acid. But after boiling this in distilled water there was no evidence of any effect, and after processing the residue he was able to recover 95 percent of his original sample.

The last test Smithson tried was boiling powdered tabasheer in a large quantity of "liquid acid of sugar" (oxalic acid). But, as before, he reported no effect and this time reported recovering 99 percent of his original sample—an impressive feat.

42

One acid that Smithson did *not* try was "fluor acid" (hydrofluoric acid), which is initially puzzling, because this was the only acid then known to affect siliceous minerals. This would seem to have been an important test, but fluor acid was extremely difficult to work with. In its liquid form it dissolves glass, which meant that in Smithson's time it could only be stored in lead or tin bottles with wax stoppers. Furthermore, as a gas it is extremely toxic and can cause immediate and permanent damage to any tissue it comes in contact with, particularly the eyes and lungs. Smithson likely concluded that it was too dangerous a material to work with.

These tests with liquid acids offered more clear evidence that tabasheer was a siliceous material, but the most convincing evidence for this would come from the tests with "liquid alkalies," which Smithson performed next.

Throughout the eighteenth century, only three kinds of alkalis were known: "vegetable" (potash), "marine" (soda), and "volatile" (ammonia), with mild and caustic forms of each. By Smithson's time it was widely known that two of these alkalis, potash and soda, would dissolve siliceous minerals. This had become a common wet test for its presence.

Accordingly, Smithson began this series of tests by boiling some tabasheer in "liquid caustic vegetable alkali" (caustic potash in water), which quickly dissolved most of it. He set the solution aside to cool, and the next morning he reported that the liquid had a slightly alkaline taste and that a drop changed the color of "tincture of dried red cabbage" green, confirming that it was alkaline. Smithson poured this solution into several containers and began to test it. He poured some into a shallow glass bowl and let it evaporate. After a day or two "it was converted into a firm, milky jelly," and over the next few days this jelly became whiter, more opaque, and finally began to dry, cracking into several pieces and curling up from the glass. Another part of the alkaline solution, this time diluted with a large amount of water, showed the same behavior, except that "the jelly was much thinner, and dried into a white powder."

Smithson did not explain this test, perhaps assuming that his readers would know what it meant, but producing a whitish jelly was a distinctive characteristic of siliceous earth in an alkaline solution. The jelly could form in two different ways, either as the solution evaporated (as Smithson described), or if an acid was added to the solution. This explains his observation that a sample of the solution "kept for many weeks in a bottle closely stopped," did not form a jelly or change in any way. It could not form a jelly if it was not exposed to the atmosphere and its water allowed to evaporate.

Some of the solution was next poured into glasses, and each was tested with either marine acid, vitriolic acid, or distilled vinegar. None of these initially seemed to have any effect, but after a few days "these mixtures changed into jellies so firm, that the glasses containing them were inverted without their falling out." This reaction took place whether the glasses were closed or open, and was not affected by light or the lack of it. Smithson reported that all the jellies produced in these tests were subsequently diluted with water, collected on a filter, and dried, and that in all cases this material "appeared to be the Tabasheer unchanged." Instead of a powder, some of the dried jellies formed small flakes, and he noted that when placed in water, these flakes became transparent and showed different colors depending on whether light was passed through or reflected off them—an effect he had described earlier with some of the samples from Hyderabad.

Finally, because distilled alcohol is very slightly alkaline, he tested the remaining solution by pouring it into a bottle of "spirit of wine" (aqueous ethanol). However, the two liquids failed to combine. The solution settled at the bottom, and when recovered and put in a shallow glass, it still formed a jelly. Tests on the alcohol showed that it did not contain any of the tabasheer.

The second alkali that Smithson tried was "strong caustic volatile alkali" (ammonium hydroxide). This was not known to affect siliceous minerals like flint or quartz, so he only used a tiny amount of his sample for this test, boiling it in the caustic liquid for "a considerable time." As expected, after being washed and dried, the weight of the sample was exactly the same.

The third alkali, "crystals of soda" (sodium carbonate), was dissolved in water and boiled along with finely powdered tabasheer for three hours. The tabasheer was not strongly affected, and so Smithson removed it, boiled it in acid, washed it in water, and then boiled it again in the alkali. After all this, the sample shrank from twenty-seven to twenty-four grains. As it cooled, the alkali solution began to form a thin jelly, which after a few days became a dense whitish jelly covered by a clear liquid. Smithson filtered it to collect the jelly, which, when dried and tested with an acid (to no effect), Smithson declared to be tabasheer. The liquid that passed through the filter was "exposed to the air in a saucer" and did produce a small amount of jelly, which Smithson removed, and he reported that as the remaining liquid evaporated, it left "regular crystals of soda," nearly identical to the soda crystals he had started with. All these tests agreed quite well with what Smithson's reference books said would happen to powdered flint or quartz when exposed to liquid alkalis.

Performing all these tests must have been tedious. But in his article's conclusion, Smithson revealed that he also performed many of these same tests on "powdered common flints" to confirm that they actually worked as the literature suggested—and in the case of one of the alkalis, they did not. Smithson reported that the flint "was scarcely at all acted upon" by liquid caustic alkali, even though it had easily dissolved tabasheer. Because Smithson's analysis depended on tabasheer's acting exactly like flint, this was a problem, but he solved it with a clever experiment. He boiled powdered flint in alkaline water to first make liquor silicum. Adding an acid to this solution produced a precipitate of pure silex, which Smithson found to "dissolve readily" in liquid caustic alkali and which, on being "exposed to the air in a shallow glass," now produced a jelly. He concluded that tabasheer and flint were made from the same siliceous earth, but in different forms, which explained their slightly different reactions with the alkali. Smithson thus anticipated a potential objection to his analysis and removed it.

The final set of experiments involved melting samples of tabasheer with other materials to make different kinds of "glass." These tests were largely based on the practices of commercial glassmaking. For example, it was common knowledge that adding "soda" (sodium carbonate) to sand (silica) lowered its melting point and that heating it produced a transparent kind of glass. Unfortunately, this "soda glass" had the inconvenient property of being soluble in water. But if "lime" (calcium oxide) was also added to the mixture, it formed "soda lime glass," which was resistant to both water and acids. By Smithson's time, a series of tests had been developed that could identify substances by the kind of glass they produced and the effects observed in the process. These tests were done with the blowpipe and typically employed "fluxes," substances like soda or borax, which would form glass with a wide range of materials.

Accordingly, Smithson's first test consisted of combining roughly equal amounts of tabasheer and soda in an attempt to make glass. For this test he used a blowpipe and, as before, a piece of charcoal to hold the sample. When he directed the flame onto the mixture he found that instead of forming a glass, "the Tabasheer quickly dissolved, and the whole spread on the coal [charcoal], soaked into it, and vanished." He had clearly used too much soda, but on his next attempt he worked more judiciously and this time, "by adding the alkali to the bit of Tabasheer in exceedingly small quantities at a time, this substance was converted into a pearl of clear colourless glass." This seemed to generally agree with Bergman's test for silica: "if the siliceous earth dissolved exceeds the weight of the flux [in this case, soda], it yields a pellucid [transparent] glass."

To complete this test, Smithson needed to recover the tabasheer from the glass, but the "pearl" he had made with the blowpipe was quite small, probably only a few millimeters across, and did not provide enough material to work with. He repeated the test, but this time with a small platinum crucible and his forge. Although not inexpensive, platinum was a good material for this kind of work, since it did not react with most materials and could withstand extremely high heat.

He combined five grains of powdered tabasheer along with a hundred grains of "crystals of soda" (sodium carbonate) in the crucible, melting it to form a white opaque glass. He removed this glass and placed it in distilled water where it completely dissolved, just as soda glass would. He then analyzed the liquid in the usual "wet" way: adding marine acid (hydrochloric acid) to the solution, which turned it into a jelly that, after being processed, yielded a powder weighing 4.5 grains and "seemed to be the Tabasheer unaltered." In his account of this test, Smithson included an anecdote that speaks to the nature of eighteenth-century chemistry, which even on this level was often conducted with everyday materials. He reported recovering "a very small quantity of a red precipitate" toward the end of this analysis, which he credited as being dye from the pink blotting paper he had used as a filter.

In the next test he changed the proportions, this time combining ten grains of tabasheer with an equal weight of soda. He again used the platinum crucible, this time heating it in a "strong fire" for fifteen minutes. This produced, as expected, a transparent glass from which, as before, he was able to recover about nine-tenths of the original tabasheer.

Earlier in the century it had been discovered that "vegetable alkali" (potassium carbonate) is an alternative to soda in the manufacture of glass, and Smithson now tried it with tabasheer. He melted equal weights of each in the platinum crucible and reported that "the glass produced was transparent; but it had a fiery taste, and soon attracted the moisture of the air, and dissolved into a thick liquor." In a follow-up experiment he changed the proportions and reported that "two parts [by weight] of vegetable alkali, with three of Tabasheer, yielded a transparent glass, which was permanent." This agreed well with one of Cronstedt's tests for silica: "If the fixt alkali [potassium carbonate] is only half the weight of the siliceous earth, it produces a diaphanous and hard glass: but when it is in a double or triple proportion, then the glass deliquesces [becomes liquid] of itself by attracting the humidity of the atmosphere."

In the last group of tests, Smithson tried combining tabasheer with a variety of other fluxes. These were his final experiments, and from the small scale on which they were conducted one gets the impression that he was using up the last bits of his sample. He performed most of these with the blowpipe. He started with the "borax bead" test, in which a small loop is made on the end of a platinum wire, heated to incandescence and then quickly dipped into powdered borax. The small amount of borax that sticks to the wire is then heated by the blowpipe to melt and form a small transparent bead. While still soft, the bead (still attached to the wire) is dipped into a tiny amount of the compound to be tested and then remelted. The color and transparency of the resulting bead identifies the material being tested.

When Smithson tested tabasheer this way, he used a single tiny fragment and reported being able to see it inside the bead, "turning about in the flux, dissolving with great difficulty and very slowly. When the solution was effected, the saline pearl remained perfectly clear and colourless." This agreed well with the Irish chemist Richard Kirwan's description of the test, where silica "gives no tinge to borax in fusion," but not as well with Cronstedt's, who reported that silica "melts easily with borax." In an age when almost no materials were perfectly pure and procedures were far from standardized, one part of the chemist's job was to deal with these kinds of inconsistencies. In this case, since some melting took place in each test, Smithson took both tests as confirming the presence of silica.

The most common flux after borax was "microcosmic salt," a complex crystalline material originally found in urine, but siliceous earth was one of the few materials that would not dissolve in it. So Smithson instead chose to use a different salt, "phosphoric ammoniac" (ammonium phosphate). Inconveniently, this compound is so unstable that Smithson had to make it himself just before he used it, but when combined with tabasheer it melted easily and formed a "white frothy bead." Smithson presumably could have adjusted the proportions to also make a transparent bead from these materials, but he believed this step to be unnecessary. At this point in the analysis he only wanted to show that nothing contradicted his overall conclusion.

In the third test, Smithson tested tabasheer with two lesser-known fluxes: vitriol of tartar (potassium bisulfate) and vitriol of soda (sodium bisulfate). Both could be used with the blowpipe to indicate the presence of the metals "alumina, magnesia, tin and zinc." Since these fluxes did not make glass, Smithson worked with them on a small platinum plate: grinding a small amount of each into a powder, mixing it with an even smaller amount of powdered tabasheer, and then melting it with his blowpipe to form a small fluid globule. Neither of these fluxes seemed to have any effect on the tabasheer, which Smithson could see "continuing to revolve in the fluid globules without sustaining any sensible diminution of size."

The fourth test involved the use of "litharge" (lead oxide), which was also a flux and when fused with quartz was known to produce a yellow glass. Litharge also had a number of important commercial uses, including the manufacture of flint glass (lead-potash glass), which was used in chandeliers and was also extremely important in precision lens making. Smithson chose to perform this experiment on a plate of silver, which has a much higher melting point than lead and would not react with it. Directing the blowpipe onto a small piece of tabasheer covered with powdered litharge, he was able to melt it "with difficulty" into a round bead. He tried working the bead on a piece of charcoal, where he could use higher temperatures, but although the bead would melt, he could not get it to become transparent. Smithson did not speculate on the reason for this, but, as with some of the other fluxes, this was probably due to not having the right proportions.

In his final three experiments, Smithson tried to melt tabasheer with other earths: first with "calcareous earth" (calcium oxide), then with "magnesia" (magnesium oxide), and finally with "earth of alum" (aluminum oxide). None of these combinations, either at the blowpipe or in the platinum crucible, showed the least inclination to form a glass. In the conclusion of his article, Smithson suggested that his experiment with earth of alum (mixed with calcium carbonate, or "whiting") failed because of either "an inaccuracy in the proportions of the earths to each other, or on a deficiency of heat." He chose not to repeat this experiment, possibly because he had no more tabasheer left to use.

In summarizing his analysis, Smithson rightly observed that his tests on tabasheer "afford the strongest reasons to consider it as perfectly identical with common *siliceous earth*." He also noted that his tests uncovered several differences between tabasheer and quartz (the purest siliceous mineral), and it was only after a detailed discussion of each of these differences that he concluded, with characteristic reserve, that none of them provided sufficient "grounds for considering it [tabasheer] as a new substance."

Although he declined to state it directly, what Smithson had discovered was a new vegetable "principle"—a new kind of plant chemistry. Before this, plant materials had been identified either by their reaction to acids, effervescing to give off carbon dioxide gas, or by the fact that they could be burned or fermented. Tabasheer did none of these things, and yet it was the product of a plant. And it was made from silica, which up to then had only been found in minerals. This is what Smithson was referring to when he stated simply that "the nature of this substance is very different from what might have been expected in the product of a vegetable."

Despite his modesty, Smithson was well aware of the significance of his work with tabasheer. In order to establish that silica was part of bamboo's essential chemistry, and not just a material that somehow formed in the plant's cavities, he now investigated the plant itself. Taking some of the bamboo "reeds" from Russell's package, he sawed away all the porous sections of the plant, leaving only the dense knots at the center of each joint. His concern was to avoid any parts that might have particles of Tabasheer "mechanically lodged in them." After burning these knots and then boiling the ashes in acid, he was able to isolate "a very large quantity of a whitish insoluble powder" that subsequent analysis showed to be silica.

Having established that silica pervaded the entire bamboo plant, he now demonstrated that it could be found in other plants as well. Repeating the test, but this time with wood ashes, he again found the whitish silica powder, although now in a much smaller proportion. Silica, it appeared, was an essential plant ingredient. Smithson had made a fundamental new discovery, and it began to generate interest even before it was published. Shortly after finishing his experiments, he and Joseph Banks, who had given him the assignment, went to inspect some bamboo growing in the hothouse of a London botanist. Banks split open a stalk with a suspicious rattle and discovered "a solid pebble, about the size of half a pea."

Smithson's report, "An Account of Some Chemical Experiments on Tabasheer," was read to the Royal Society on July 7, 1791, almost exactly one year after he began working on it. It was quickly published in the *Philosophical Transactions*, and it soon became apparent that Banks was not the only one impressed with Smithson's analysis. Notices about the article and its surprising findings began to appear in a wide range of journals. The *Monthly Review* noted "the experiments of Mr. Macie [Smithson], very judiciously executed," and the *Critical Review* remarked that "this is a discovery, we believe, wholly new; for the earth of vegetables . . . was supposed to be exclusively lime." The *English Review* made a similar observation: "this singular vegetable concretion is of a siliceous nature, and not calcareous, as might have been expected." A long summary of the article appeared in the journal *Medical Facts and Observations*. The article also attracted the attention of foreign journals. A notice about it appeared in the *Annales de Chimie* and the entire article was reprinted, with commentary, in the *Journal de Physique* in 1792. In time, notices about Smithson's article appeared in most of the major European journals.

Smithson's discovery was also of great interest to botanists, who struggled to find a place for it in their organizational systems. As James Edward Smith, the distinguished president of the Linnean Society, would write, "Nothing is more astonishing than the secretion of flinty earth by plants, which, though never suspected till within a few years, appears to me well ascertained." Smithson's article came to the attention of another important scientific figure, the Prussian explorer-naturalist Alexander von Humboldt. In the fall of 1802, Humboldt was in the mountains of Peru when he came across some bamboo plants with a suspicious rattle, and he later wrote to his colleagues back in Paris: "We shall bring with us also a siliceous substance analogous to the tabascher of the East Indies, which M. Macé has analysed."

True to his word, when Humboldt finally returned to Paris in 1804, one of the many materials he brought back for analysis was the South American tabasheer. The actual analysis was performed by two distinguished French chemists, Antoine Fourcroy and Nicolas-Louis Vauquelin, and their investigation, like Smithson's, showed it to be composed mostly of silica. But they reported that only about 70 percent of their sample was siliceous and that the rest of its weight came from "potash, lime and water," which seemed to call Smithson's analysis into question. The question of which analysis was correct, or whether there was a fundamental difference between American and Indian tabasheer, remained unresolved for many years. English calls for a reexamination of the Peruvian tabasheer went unanswered; given that the Napoleonic Wars were still raging during this period, the issue may have become entangled with national pride.

The question was never formally resolved, but in 1828 Edward Turner, professor of chemistry at London University, conducted his own analysis of tabasheer and found it to agree with Smithson's. And in 1837 the eminent Scottish chemist Thomas Thomson effectively ended the discussion when he praised the "very minute, accurate, and complete set of experiments, by this acute and accomplished philosopher [Smithson]" and then wrote, "In the year 1806, a specimen of tabasheer, from Peru, was put into the hands of Fourcroy and Vauquelin, by Humboldt and Bompland [Humboldt's fellow explorer]. These chemists subjected it to analysis, extracted from it 70 per cent of silica, together with a little lime, and concluded (though it is not easy to see the evidence) that the tabasheer, which they examined, was a compound of 70 parts of silica, and 30 parts of potash. But under the potash were included the vegetable matter which they showed it to contain, and also, the water, the amount of which, they seem not to have thought of determining." In the clinical prose of natural science, this was scathing criticism. It effectively destroyed the credibility of the French analysis and ended any question about the accuracy of Smithson's analysis.

Perhaps the most surprising use of Smithson's analysis was its application to geology. Studies of the Earth were still dominated by speculation about how it had formed, and each theory looked to new scientific discoveries for support. Smithson's friend Richard Kirwan, for example, seized on the fact that silica could be soluble in water to support the "neptunist" theory that the Earth's solid materials all had their origin in a primordial solution. Another geological theory being discussed in the late eighteenth century was that of the French naturalist Baptiste Lamarck, who held that Earth's crust, or at least its upper layers, had been formed by the actions (and from the remains) of plants and animals. The idea had much to recommend it, not the least being that it offered an explanation for the existence of animal and plant fossils, and Smithson's study seemed to support it. In 1805 an item in the *Philosophical Magazine* remarked: "Do animals form lime, and vegetables argil and silex, as some naturalists assert? The generation of stones and that of mountains, and the whole history of our globe, depend in some measure on this problem. It is to it we may refer the analysis of the tabasheer."

Although tabasheer's relevance to geology would eventually be dismissed, this kind of speculation ensured that Smithson's article would continue to be discussed long after it appeared in the scientific literature. Smithson was only twenty-six when his article was published, and, despite its rather dense prose, his analysis was an impressive technical achievement. He seemed to anticipate and answer every possible objection, and neither his work nor his conclusions were ever seriously questioned. The fact that Smithson was even given this project speaks to the way he was already regarded within the Royal Society, and the way he responded to this opportunity gave him an international reputation and put him in the front rank of English chemists. The potential that everyone had seen in him was beginning to be realized.

———————†———————

4.

CALAMINE

SMITHSON'S 1791 ARTICLE on tabasheer was by any measure a great success. It burnished his reputation in England and quickly brought him to the attention of the international scientific community. But his success was tempered by a personal tragedy that took place at almost the same time: the ruin of William Thomson.

In the fall of 1790, at the same time that Smithson would have been conducting his tabasheer experiments, Thomson suffered what he described as "a most scandalous imputation." Few details are known, but on September 14, an inquiry took place at Oxford concerning "a Charge of an unnatural kind against Dr William Thomson." The use of "unnatural" in this context was a reference to homosexuality, which was illegal and considered immoral in eighteenth-century England. When the university's governing body subsequently charged him with suspicion of "sodomy and other unnatural and detestable practices with a servant boy," Thomson was in serious trouble.

It had been about seven years since he and Smithson first met, and during those years Thomson's rise in English science had been as impressive as Smithson's. Thomson had been elected a member of the Royal Society and was the anatomy lecturer at Oxford as well as a physician at Oxford's Radcliffe Infirmary. Throughout this time Thomson was an important figure in Smithson's life.

The event being investigated at Oxford had actually taken place four years earlier, and Thomson explained it to his friends as a medical experiment—a defense made plausible by the fact that he was a physician and the university's lecturer on anatomy. This, combined with the fact that some of Thomson's colleagues were willing to testify in his support, makes it seem that he could have mounted a successful defense. But he needed to act quickly and decisively, and when he failed to even respond to the charges, his fate was set. Within two weeks of being charged he was reported to have left the university, and on November 2, 1790, the vice chancellor made it official. Thomson was stripped of his positions and his degrees and forever banished from the university.

Thomson was almost certainly in London when this happened and probably making hurried preparations to move to the continent. One of the consequences of "the ruin of his moral character," as one of his friends put it, was that Thomson was now widely seen as socially unfit, and if he remained in England he could have been subject to arrest. He would never again set foot on English soil, but as he prepared to leave there was still one last painful task to which he felt obliged to attend.

To keep his disgrace from being associated with any of his scientific colleagues, Thomson sent letters of resignation to the Royal Society and all the other scientific organizations that had been such a significant part of his life. These letters, which were read at each organization's next meeting, had the effect of ensuring that his situation soon became widely known. Even before his name was officially stricken from each organization's membership list—as it quickly was—Thomson's scientific life, at least in England, was effectively over.

Once he was safely on the continent, Thomson stayed briefly in Paris before making his way south to Italy. He changed his name from William to Guglielmo Thomson and joined the small colony of British citizens living in Naples. There, literally in the shadow of Mount Vesuvius, he began to practice medicine again and, being a skilled physician, soon had a thriving practice. The Pope himself is said to have used his services.

Although continuing to associate with Thomson put his own reputation at risk, at least in England, Smithson made plans to visit him in Italy. He left London around November 1791, shortly after his tabasheer article was published, and, as he gave up his London residence before he left, he may have intended to leave England permanently. But he only got as far as Paris before his plans changed. Once there, he almost immediately postponed his visit to Italy, writing to a friend, "It is now my intention to spend the *next* winter in Italy, possibly at Naples."

It was an understandable decision. Like many of his contemporaries, Smithson had a sense of living at a pivotal time in history, and the startling developments of the French Revolution certainly confirmed that perception. Most Englishmen viewed the Revolution with a certain amount of trepidation, but for Smithson and many of his friends, the unfolding spectacle that followed the Revolution's initial violence appeared to be the triumph of reason and—just possibly—the beginning of a transformation of human society. Like many of his age and class, Smithson was anxious to witness these historic events in person, and he explained his decision to remain in Paris with an apt metaphor: "I should like much to see the lava which is at present running from Vesuvius: however I console myself in some degree for the loss of that sight by the consideration that I am here on the brink of the crater of a great volcano, from whence lavas are daily issuing, but whose effects are widely different from those of the other, that is laying waste one of the finest countries in the world; is threatening with ruin the noblest efforts of human art; while this, on the contrary, is consolidating the throne of justice and reason; pours its Destruction only on erroneous or corrupt institutions; over-throwing not fine statues and amphitheaters, but monks and convents."

Smithson's idealism was palpable, and it may even have sounded radical. But in late eighteenth-century England, an interest in science often lived alongside radical political views—"radical" in this context being characterized by two positions: favoring democracy over monarchy, and deism (or, more radically, atheism) over conventional religion. Smithson's letters from Paris reveal his views on both these topics.

With regard to the monarchy, Smithson was an early and enthusiastic supporter of the French Revolution, which he described as "consolidating the throne of justice and reason," but not under a king. He wrote to his friends about the great "inconvenience" of having a monarch and dismissively referred to the king of France as "Mr. Louis Bourbon": "Mr. Louis Bourbon is still at Paris, and the office of king is not yet abolished, but they daily feel the inutility, or rather great inconvenience, of continuing it, and its duration will probably not be long. May other nations, at the time of their reforms, be wise enough to cast off, at first, the contemptible incumbrance. I consider a nation with a king as a man who takes a lion as a guard-dog—if he knocks out his teeth he renders him useless, while if he leaves the lion [with] his teeth the lion eats him."

Smithson expressed his views on the church in his metaphor of the great volcano of the Revolution, which "pours its Destruction only on erroneous or corrupt institutions." A few months later he reported, with no expression of regret, that "the church is now here quite unacknowledged by the state, and is indeed allowed to exist only till they have leisure to give it the final death stroke."

Smithson was twenty-seven when he wrote these letters, and subsequent events would soon make him cynical about the Revolution, but his fundamental idealism seems to have remained intact. Nine years later he would still write enthusiastically about "the change which, in the space of a few short years, has taken place, throughout all Europe, in the minds of men" and about the "liberality of sentiment" that was now "tending to the union of all men into one nation, and the conversion of all religions into their amiable and evident principles of inoffensiveness essential to the existence of men in a collected state."

Despite the awkward prose, these are the sentiments of Deism, the high Enlightenment's "religion of reason." Deism rejected the teachings of established religion, which it saw as being based solely on miracles and revelation, and instead attempted to arrive at religious truth through the application of observation and reason. The attraction of this approach to a man like Smithson is not hard to see. Deism applied much the same methodology to understanding the spiritual world that science used to explore the physical one.

Living in Paris afforded not only proximity to the spectacle of the Revolution but also access to the French scientific community, and Smithson lingered there for nearly a year, not leaving until September 1792, just days before the French monarchy was officially abolished. Events were becoming increasingly bloody, and Paris had become dangerous, especially for foreigners, most of whom had already left. On January 21, 1793, King Louis XVI was executed.

By that time Smithson was on his way to Italy. He chose to travel east through the Tyrol region in the Alps before heading south, which prudently allowed him to avoid areas with military activity and also gave him a chance to indulge in some mineral collecting. During his time in Paris, Smithson had continued to pursue his scientific activities, and each of the letters in which he wrote so passionately about the Revolution also contained requests that his friends send him crystals and scientific information. Smithson probably also traveled with a portable chemical laboratory, which would have enabled him to perform basic chemical analysis anywhere he went. Surprisingly complete, with portable scientific instruments and a selection of chemical reagents, these sets were commonly carried by traveling naturalists and were readily available in the scientific instrument shops of either London or Paris. They typically came in one or two fitted wooden cabinets about the size of a small suitcase.

By July 1793, Smithson was in Florence, where he was well received by the local scientific community. His wealth allowed him to live wherever he wanted, and so Florence became his base of operations for the next three years as he made a series of side trips to explore Italy. The first of these excursions began along Italy's west coast, with Smithson traveling south toward Rome and Naples. He made frequent stops along the way, collecting mineral samples and making observations, and he spent some time in Rome, which held little scientific interest for him. Then he almost certainly continued south to Naples to see Vesuvius and, presumably, Thomson. No record of that part of the trip survives, however. What is known is that the two men continued to correspond and, in particular, that in May 1794 Thomson sent Smithson a package of special mineral samples from Mount Vesuvius.

Living in Naples had provided Thomson an opportunity to collect samples of Vesuvius's many eruptions, and this included a whitish "saline substance" that had flowed from one of its vents. He included it in a package to Smithson, along with the request that Smithson analyze it, which he quickly did. He found that it contained "vitriolated tartar" (potassium sulfate), a material well known to chemists but never before associated with volcanoes. Recognizing the importance of this discovery, Thomson mentioned it in a thirty-five-page Italian-language brochure he wrote about his collection, carefully giving Smithson credit for the identification. However, Thomson's uncertain social status made other savants reluctant to engage with him, and even after being translated into French, the publication went unnoticed. Smithson carried the rest of the sample back to England, and more than a decade later he published a new, more complete analysis of it.

Smithson appears to have been comfortable in Italy, and he only left after French forces, under a young commander named Napoleon, suddenly captured Piedmont in early 1796. Smithson was north of Milan at the time, and he prudently moved further north into Austria. This put distance between him and any potential military actions, but it also made his return to Italy problematic. After visiting the historic lead and zinc mines near modern Bleiberg (pausing, of course, to collect samples), he continued north to Germany. His exact itinerary is not known, but by May 1796 he and his servant were in Dresden, which had a large English community. He spent several months touring Germany, visiting the famous mines of Saxony and the equally famous mining academy at Freiberg, which was the intellectual center of the neptunist theory of geology. He also spent time in Berlin and became friends with Martin Klaproth, perhaps the foremost German chemist of his time.

Because the French controlled all the ports on their side of the English Channel, Smithson's only option for returning to England was to sail from northern Germany across the North Sea. It was a rough passage, but by June 1797 Smithson was finally back in London. He had been gone for almost six years. England was full of fears of a French invasion, and the future seemed far from certain, but for Smithson personally this may have been a golden time. He was still only thirty-two and wealthy, with a rich intellectual life and important and interesting friends.

He quickly reengaged with the English scientific community. In 1799 he was invited to become one of the founders of the Royal Institution, contributing the requested sum of fifty guineas to become a "proprietor" for life. The institution, led after 1801 by Humphry Davy, would soon become one of England's scientific centers. Meanwhile, in January 1801, Smithson was sworn in as a member of the Royal Society's council, its governing body. There he joined Cavendish and several of his other friends. Smithson was now an insider in the English scientific community.

But on March 15, 1800, in the midst of all this public recognition, Smithson's mother died. The loss affected him strongly, and he seems to have withdrawn from social contact for several months. His mother's death also meant that he needed to untangle her financial and legal affairs, which was a daunting task—and certainly required time. This was also the time that he chose to honor his mother's request that he legally change his name from "Macie" to "Smithson." After months of waiting, the request was officially granted in February 1801. He was now James Smithson.

These obligations monopolized Smithson's attention for most of 1801, and this probably explains why, after being sworn in as a member of the Royal Society's council, he waited until the following year to attend his first meeting. Under the circumstances, he could have been excused for putting his scientific work on hold, but in 1802 he showed that this had not been the case.

On November 18, 1802, Smithson's second scientific paper was read to the Royal Society. Titled simply "A Chemical Analysis of Some Calamines," it was a complex and remarkable work. But it was also part of a much larger story, a story that starts with brass, one of the most versatile and useful materials of Smithson's time. It was relatively cheap, easy to work with, durable, and seemed to be used for everything—but it could be tricky to make. Brass is an alloy of two separate metals, copper and zinc, which fuse together when heated in a furnace. Copper is easy to produce—rocks containing it are strongly heated until liquid copper flows out. By itself, copper is too soft for most applications, but when it is combined with zinc, the resulting brass is much harder, stronger, and more corrosion-resistant than either metal by itself.

Brass has been made in limited quantities since antiquity, and the first examples were probably made by accident from rocks that just happened to contain both copper and zinc. But it was a difficult process to control, because of zinc's unusual sensitivity to heat. Copper melts at around 1,100°C and becomes a gas at around 2,600°C, whereas zinc melts at around 420°C and turns into a gas at around 900°C.

The first page of Smithson's original "Calamines" manuscript, written in his own hand. This is what was read at the Royal Society meeting of November 18, 1802, and later sent to the printer for use in the *Philosophical Transactions*. The crossed-out words and substitutions are the editor's corrections. ©The Royal Society.

This is very low for a metal and much lower than the melting point of copper. If copper and zinc (or rocks containing zinc) are placed in an open crucible and heated together, the zinc will evaporate and be lost in the atmosphere before the copper even melts—and no brass will be produced.

It is believed that the Romans finally solved this problem by simply making their brass in covered ceramic pots. Sealed with clay (to keep the zinc gas from escaping) and containing alternating layers of copper, zinc ore, and charcoal, the large pots needed to be kept at a high temperature in coal-fired furnaces for many hours. But once they cooled enough to be opened, the copper inside was found to have been converted into brass, and the rest of the material fused into a slag that could be discarded. Making brass this way was a tedious but workable solution, and over time it proved to be a reliable method of making sizable amounts of reasonably good brass. But it was all based on trial and error, with little understanding of what transpired inside the sealed pot.

The utility of brass meant that mines that produced high quality zinc ores were extremely valuable. As far back as the Middle Ages, most of the zinc in Western Europe (as well as most of the brass) came from mines around the village of Kelmis, in what is today eastern Belgium. Historically known by the French name La Calamine, the area became so identified with the production of brass that any rock containing zinc came to be called "calamine," and any brass made in pots with zinc-bearing rocks was called "calamine brass."

By the sixteenth century, brass had become so important that its manufacture was a national priority. When Elizabeth I ascended the throne in 1558, England had essentially no domestic brass industry, and the country's reliance on imported brass was seen as a strategic disadvantage. Accordingly, in 1568 the queen issued royal charters to two new companies, giving them broad monopolies together with some very specific responsibilities. The first company was the Society of Mines Royal, which was to develop the capacity to produce English copper, and the second was the Company of Mineral and Battery Works, whose charge was first to find and develop domestic sources of zinc, and then to establish an English brass industry.

Fulfilling these charters meant finding deposits of both copper and zinc on English soil, and while the search for copper enjoyed quick success, finding zinc deposits initially proved elusive. Eventually a usable zinc deposit was found in Somerset, which allowed the production of English brass to begin, but for decades the company continued to search for additional sources of the important metal. The search for zinc even extended to the New World, and scholars have found that the Jamestown colony (1607) in Virginia was partly organized for just this purpose. The Mines Royal and the Mineral and Battery Works both provided important financial support for the colony, and analysis of crucibles and metallic residues found on the Jamestown site have offered convincing evidence that local minerals were being tested for their ability to produce brass. The Jamestown settlers never found any calamine deposits, but their search demonstrates that calamine has long been a topic of special interest in England.

In Smithson's time zinc's properties were still of considerable scientific interest. Residues in crucibles recovered from the Oxford chemistry lab and dating from just before the time of Smithson's enrollment show clear evidence of experimentation with calamine. Martin Wall, Smithson's chemistry instructor, lectured on zinc in his chemistry course, addressing speculations that it was a "semi-metal," as opposed to "perfect" metals like gold and silver. Calamine and brass were also topics of discussion at meetings of the Coffee House Philosophical Society in London, which Smithson attended. On one occasion a member brought an unusual item: "a piece of Brass foil remarkable for its not being disposed to tarnish by long keeping. He affirmed that in the manufacturing it had been rolled in the form of copper & afterwards converted to brass by Fumigation with Calamine."

Despite the importance of zinc, the long history of its use, and all the effort that had been expended to understand it, little progress had been made in understanding the ores that contained it. Through the seventeenth century, "calamine" had referred to any rock that contained a significant amount of zinc. But some of these rocks worked better for making brass than others, and by the eighteenth century, miners and brass manufacturers had learned to select a particular zinc ore, which they called "blende" (zinc sulfide), for making brass. It was the first individual zinc ore to be identified, and it not only worked well for making brass but was also fairly easy to distinguish from the many other types of rocks found in zinc mines. Unfortunately, an understanding of the other zinc ores remained elusive, and the term calamine was now applied to any mineral containing zinc, except blende.

In 1782 the French chemist Bertrand Pelletier had reported a second zinc ore, zinc silicate, which he identified simply as *pierre calaminaire*—calamine. As Smithson began his study, this analysis had yet to be confirmed, and there was a suspicion among mineralogists that even more zinc ores were waiting to be found.

After centuries of study and use, why was there so much confusion about these minerals? The problem stems from the way in which zinc deposits are formed. It is now understood that these deposits are created when geological events force water from deep underground up through cracks and spaces in the existing layers of rock. The water, ranging up to 150°C and under considerable pressure, brings with it a variety of dissolved substances including zinc, lead, iron, sulfur, copper, cadmium, silver, and arsenic. Over time, as this water cools and slowly withdraws, a layer of these materials will be left behind, forming deposits that line the exposed surfaces of any cracks or chambers that the water was able to reach. As this cycle repeats itself over millions of years, these materials begin to react chemically both with each other and with the existing bedrock. The resulting jumble of different minerals, rarely pure and often so intermixed that they could hardly be distinguished, had defied analysis.

Thus, as Smithson began his study of calamine, he addressed not only an important commercial question but also one of the oldest and most intractable problems in all of mineralogy. It had been partially solved by the identification of "blende," and by the tentative identification of a zinc silicate, but as he noted at the very beginning of his article, when it came to the ores of zinc, "their constitution was far from decided, nor was it even determined whether all calamines were of the same species, or whether there were several kinds of them." These were some of the questions that Smithson set out to explore.

It had been eleven years since his last article, and Smithson was now thirty-seven. His writing style had changed over the years; his tone was now more confident and assertive, and his scientific style had changed as well. Long descriptions of the sample's physical characteristics were gone, as were the tedious accounts of every single test. Smithson now reported only those tests that were necessary to identify his samples.

Most of the tests he used were similar to those in his tabasheer article—specific density, reaction to acids and alkalis, and frequent measurements of weight—but there were some interesting variations as well. For instance, Smithson made no mention of using either a furnace or a forge in this article; all his tests with heat and flame were now made exclusively with the blowpipe. Where he formerly had needed a forge (or furnace) with a raging fire to heat his samples, now he could accomplish the same thing sitting at a table, using just a candle (or lamp), a blowpipe, and his skill. In addition to being more convenient, this method allowed him to use much smaller samples in his tests, which was a significant advantage.

Smithson used the blowpipe's flame to reveal the presence of zinc in his samples, and he described this test in some detail, but with prose that challenges the modern reader: "Urged with the blue flame, it [the sample] became extremely friable; spread yellow flowers on the coal; and, on continuing the fire no very long time, entirely exhaled. If the flame was directed against the flowers, which had settled on the coal, they shone with a vivid light." What Smithson was describing was a process that began with him using the blowpipe to heat a small sample of calamine. Because of the heat involved, he placed it in a depression in a piece of "coal" (charcoal), which he used as a holder. The "blue flame" was the blowpipe's most intense heat. Under these conditions it only took a few seconds for the sample to get so hot that it began to fall apart and to release its zinc invisibly into the air. The zinc almost immediately combined with oxygen to form zinc oxide, and then quickly settled back down on the coal as a ring of pale yellow powder that turned white as it cooled. This appearance was well known in the brass industry, where white encrustations called "zinc flowers" would collect on any surfaces exposed to zinc fumes.

As Smithson continued to heat the sample, he reported, it soon fell apart completely and disappeared. (He would later identify this effect as characteristic of zinc carbonate.) Finally, to prove that his "flowers" were really zinc, Smithson directed the flame at them. One of the characteristics of zinc oxide is that it glows yellow when heated, and he reported that his "flowers" did exactly that.

Smithson used this and many other tests to analyze a total of four calamine samples, two from English zinc mines, in Derbyshire and Somersetshire, and two from zinc mines in southern Austria, in Bleyberg and Regbania. Except for the calamine from Bleyberg, which he purchased from a dealer, he had collected all the specimens himself.

In the introduction to his article, Smithson noted that the chemist Torbern Bergman and the crystallographer René Just Haüy had already declared that there was no zinc mineral containing "carbonic acid"—a reference to the carbonate ion, CO_3. This seems to be a surprising claim, as the carbonates—those minerals containing carbonic acid—form one of the largest mineral groups. Even more curious was that many calamines seemed to contain carbonic acid when tested for it.

Calamine ores are hardly ever found in a pure state, and so the problem of analyzing them was not just determining what elements were present in the sample. It was also necessary to decide which of those elements were impurities and should be ignored. Making the wrong decisions could lead to spectacular mistakes, and this is what had happened to Bergman and Haüy. They had detected carbonic acid in their samples but dismissed it because they assumed it was an impurity. It was an understandable assumption, since the great majority of zinc deposits form on limestone, which contains carbonic acid. But it was not an assumption that Smithson made.

In his own analysis, Smithson was careful to dismiss only trace materials—those, as he put it, that were present in "too small quantity to deserve notice." His main method of dealing with impurities was to try to avoid them, which he did by choosing his samples very carefully. With the exception of the mineral from Bleyberg, which was not in a crystalline form, all the calamine samples that Smithson analyzed were small crystals that he found inside the original sample and had selected for their "perfect regularity and transparency," which "make it impossible to suppose . . . a foreign admixture in them." Smithson's ability to work with tiny samples served him well here, as these crystals were small and few in number.

Smithson's analysis proceeded smoothly, and at the end he was able to show that three of his samples contained significant amounts of carbonic acid and were, in fact, "carbonate of zinc." However, while two of the samples were nearly identical in composition, the third carbonate, the one from Bleyberg, was quite different. It contained a higher proportion of zinc, a lower proportion of carbonic acid, and a significant amount of water that seemed to be part of the mineral's chemistry and not an impurity. It was a puzzling substance and one that neither chemistry nor mineralogy could explain.

Smithson's analysis of the fourth calamine sample, from Regbania, showed that it contained a large amount of quartz (silica), which confirmed Pelletier's earlier analysis. Smithson called it "Electric Calamine" because of the strong static-electric charge it developed when heated. He was also the first to be able to measure a crystal of this mineral, and he was the first to note the striking green light it emitted when heated—which provided an easy way to identify it.

At the midpoint of his article, as he finished describing his chemical analysis, Smithson had discovered one new zinc ore (zinc carbonate) and confirmed the existence of another (zinc silicate). This was an important contribution to a historic scientific problem. It was also commercially important, and as one reads the article there is an expectation that at some point Smithson will make a straightforward statement about what he has discovered and what it might mean. He does not. Smithson rushes through the description of his analysis without so much as a sentence of summary, as though he were in a hurry to get to the next section.

That may have been the case. The investigation in the first half of Smithson's article was performed with both skill and intelligence, but it was essentially a conventional late eighteenth-century chemical analysis. It provided the identity of the materials in each of his samples, along with a measurement of their relative weights, but that was all. In the preceding decades, chemists had collected large amounts of this kind of information, and there had always been an expectation that it would, at some point, begin to reveal patterns or suggest deeper truths. But as the eighteenth century drew to a close, there was a growing realization among chemists that these insights were not being generated and that something else needed to be tried. It was in this spirit that Smithson introduced the second half of his article, describing a method he had developed to "correct" the measurements of chemical analysis with "theory."

Smithson was interested in a promising new chemical theory called "definite proportions," which argued that chemical substances always combine in specific ratios. The main champion of this theory was the French chemist Joseph Proust, who in 1799 wrote an article on copper oxides that Smithson appears to have read. Proust had been experimenting to see how much carbonic acid could combine with a given amount of copper and had discovered that it was always the same amount, the same proportion by weight. These and other experiments led him to

	Bleyberg	Somersetshire	Derbyshire
"Calx of zinc" (ZnO)	0.714	0.648	0.652
"Carbonic acid" (CO₃)	0.135	0.352	0.348
Water	0.151	0.0	0.0
Total	**1.000**	**1.000**	**1.000**
Specific Gravity	3.584/3.598	4.336	4.333

Summary of Smithson's analysis of the three calamines containing "carbonic acid," giving the proportions of their elements by weight. Note that Smithson converted the actual measurements to percentage by weight, to make them easier to compare.

the generalization that chemical substances can be characterized as having their elements combined in the same fixed proportions, by weight. This concept, in turn, led to the idea of a "combining weight": the greatest amount, by weight, of one substance that will combine with a given weight of another substance. This seemingly simple measurement, for the first time, provided chemists with an upper limit on the proportions of materials in a given substance.

This idea that the chemical composition of substances was "fixed" corresponded well with another theory being widely discussed at the time: Haüy's theory of crystallography, which seemed to be a model for other types of scientific inquiry. Smithson adopted a number of Haüy's assumptions about crystals in his new approach to chemistry, including that of "mathematical simplicity." He stipulated that the ratios in which substances combined needed to be "of very low denominators. Possibly in few cases exceeding five." He justified his belief in these "simple ratios" on the basis of "the simplicity found in all those parts of nature which are sufficiently known to discover it." Belief in the fundamental mathematical simplicity of nature was widespread in late eighteenth-century science and would continue well into the nineteenth century.

Smithson also accepted Haüy's insistence that all chemical combination is "binary"—that matter only combines in units of two. Haüy saw these binary units as the discrete physical building blocks from which crystals are formed, and Smithson argued that chemicals combined in the same way: one to one. In 1802, when Smithson wrote his article, chemists were just beginning to think about these topics, and his assumption of "binary" chemical combination initially produced no objections. Unfortunately, mineralogists would soon begin to find substances that seemed to be exceptions to Smithson's rule, but those discoveries were still several years away, and, having briefly described his theory, Smithson now turned to applying it. What he most wanted to understand was the surprising composition of the calamine from Bleyburg, which was clearly a zinc carbonate but of an entirely new form. The two other zinc carbonate samples that Smithson analyzed had been nearly identical to each other, with the same elements, in the same proportions, and with the same specific gravity. This strongly indicated that this was the "definite proportion" in which the two substances—calx of zinc and carbonic acid—ideally combined, and Smithson accordingly "corrected" his measurements, reducing them into simple ratios.

	Somersetshire	Derbyshire	Simple Ratios by Weight ("corrected by theory")
Calx of zinc (Zinc + Oxygen)	0.648	0.652	2/3
Carbonic acid (Carbon + Oxygen)	0.352	0.348	1/3
Total	**1.000**	**1.000**	

Composition of the zinc carbonates, reduced to simple ratios according to Smithson's theory.

This table, compiled from Smithson's data, shows how his assumption that all matter is "binary" worked. First, according to the binary requirement, zinc carbonate must be a combination of two "secondary" components. He could demonstrate that it was by heating a sample until carbonic acid was released as a gas and all that was left was the calx of zinc. Both of these components were also binary: calx of zinc was zinc + oxygen, and carbonic acid was carbon + oxygen. When Smithson described all these substances as "binary," he seemed to be correct.

By reducing his data to simple ratios, as called for by his theory, Smithson was able to determine each mineral's ideal "combining weight." Thus, under ideal circumstances, a given weight of carbonic acid would combine with exactly twice its weight of calx of zinc.

But Smithson found that the calamine from Bleyberg contained (by weight):

Calx of zinc	0.714
Carbonic acid	0.135
Water	0.151
	1.000

By his own calculation, that amount of carbonic acid was only sufficient to combine with part of the calx of zinc in the sample (0.135 units of carbonic acid could only combine with 0.270 units of calx of zinc, which left 0.444 units of the calx to be accounted for). In that case, since water was the only other substance left in the mineral, Smithson argued that this was what the remaining calx must have combined with. It formed what Smithson now called a "hydrate of zinc"–a new mineral that he assumed to exist because it was required by his theory. While he admitted that "hydrate of zinc" might never be found by itself in nature, he noted that the weight of the water in it compared to the weight of the remaining calx of zinc was 1 to 4, a simple ratio.

Using his "corrected" data, and reasoning from theory, Smithson presented a strong case that the Bleyberg calamine was actually a "compound salt," a chemical combination of two other minerals: carbonate of zinc and the new "hydrate of zinc." He provided an illustration of how such a compound mineral might be composed.

Today the hydrate form of zinc that Smithson studied is called hydrozincite, and its composition–$Zn_5(CO_3)_2(OH)_6$–is thought to be significantly different from what Smithson described. But when it was published, Smithson's goal was to show that his theory explained all the known facts, and in this he was successful.

1000 parts of the compound salt of carbonate and hydrate of zinc consist of,

$$
\text{Carbonate of zinc } 400 =
\begin{cases}
\text{Carbonic acid} = \dfrac{400}{3} = \quad - \quad - \quad -133\frac{1}{3} \\[2ex]
\text{Calx of zinc} = \dfrac{400 \times 2}{3} = 266\frac{2}{3}
\end{cases}
$$

$$
\text{Hydrate of zinc } = 600
\begin{cases}
\text{Calx of zinc} = \dfrac{600 \times 3}{4} = 450 \\[2ex]
\text{Ice } - - = \dfrac{600}{4} = \quad - \quad - \quad 150
\end{cases}
$$

$$= -716\frac{2}{3}$$

$$\overline{\qquad\qquad}$$
$$1000.$$

Smithson's chemical notation for the Bleyberg calamine, as predicted by his theory, shows the two minerals that combined to make his sample on the left, and the two substances that make each of them, along with their ideal weights in the center. The close agreement between the "theoretical" numbers in the right column and the results of his analysis of the Bleyberg sample is striking. Courtesy of the Smithsonian Libraries.

He was also successful in demonstrating simple ratios in the other calamines he analyzed. Looking back over Smithson's article, it is clear that his purpose in writing it was to showcase his new theory, and that the calamine samples he analyzed had been carefully selected to demonstrate its potential. He had chosen a difficult and historic problem—calamine—and with the help of his theory had resolved it. It was an important contribution to science and a strong argument in favor of the theory itself.

In his conclusion, Smithson expanded on what it might all mean. "If the theory here advanced has any foundation in truth," he wrote, "the discovery will introduce a degree of rigorous accuracy and certainty into chemistry, of which this science was thought to be ever incapable." Once it was determined, "a certain knowledge of the exact proportions of the constituent principles of bodies, may likewise open to our view harmonious analogies between the constitutions of related objects, general laws, etc. which at present totally escape us." History would soon prove him right. As one historian of science would later write, Smithson's words "might almost be taken for a prophetic description" of the development of atomic chemical theory that was just beginning in England.

Smithson's article was read to the Royal Society in November 1802, and the first review of it appeared a few weeks later. It was written by Humphry Davy, who was now at the Royal Institution. Like Smithson, he was a "philosophic" chemist, and he enthusiastically reported on the paper before it was even published, writing his long and generally positive review of it solely on the basis of having heard it read.

Smithson's article was quickly reprinted in two other journals. One of them was William Nicholson's *Journal of Natural Philosophy*, which published it in early 1803. Smithson would have remembered Nicholson from his early days in London, when they were both members of the Coffee House Philosophical Society, and Nicholson was careful to reprint the article exactly as it had been published. But he did replace Smithson's drawing of the "Electric Calamine" crystal with a more detailed engraving of it, and he added summarizing comments in the article's margins. Perhaps the most interesting of these comments was his characterization of nearly the entire second half of Smithson's article as "Elucidations of theory." Smithson's article was also republished in the 1804 *Repertory of Arts, Manufactures, and Agriculture*, a monthly journal that regularly reprinted scientific articles of special interest. The interest in this case related to the production of brass.

"Calamines" was a more challenging work than Smithson's first paper, and most of the reviews, like the one in *The Monthly Review*, simply reported the facts of the article without attempting further comment. The least enthusiast assessment appeared in *The Critical Review*. While praising the successful chemical analysis, the unnamed reviewer suggested that "Mr. Smithson refines perhaps too far, when he would draw conclusions from so few observations" and "Mr. Smithson attempts to support the theory . . . by particular remarks on the proportion of the component parts of the calamines analyzed in this paper"—suggesting a certain unease with Smithson's use of his own analysis to support his new theory. This was a criticism that Smithson took seriously and would address in a later publication.

The French reaction to Smithson's article was more muted, probably because it was published in 1803, just as the war with England was resuming. Any copies reaching France thus needed to be brought through, or around, the English blockade, and it seems not many were. Shortly after it was published, Smithson sent a copy of his article to J. C. Delamétherie, editor of the prestigious *Journal de Physique*, but Delamétherie never received it. As a result, no French translation of Smithson's complete article was ever published.

French crystallographers knew about the calamine article (Smithson would have sent a copy to his friend Haüy), and they were quick to include the new minerals he had identified in their mineralogy texts. One of the earliest was Alexandre Brongniart's *Traité élémentaire de minéralogie* (1807), which gave lengthy descriptions of both zinc carbonate and zinc silicate and acknowledged "M. J. Smithson" (Monsieur James Smithson) for their discovery. But Brongniart declined to use Smithson's term "zinc hydrate" and only mentioned the analysis of "Le zinc carbonate de Carinthie"—the sample from Bleyberg. Haüy also included Smithson's findings in the next edition of his comprehensive *Tableau*, all properly credited, but he too declined to endorse the idea of a zinc hydrate.

Zinc was becoming an important topic to the French, because as the war between England and France intensified, the subsequent mobilization of forces exposed a French strategic disadvantage—a chronic shortage of metals. Napoleon took a number of steps to address this problem, including, in 1805, granting a monopoly to a Belgian industrialist named Dony to develop the historic zinc mines around La Calamine. Dony had invented a process for producing metallic zinc that promised not only to increase the production and quality of brass but also to make large quantities of zinc available for other purposes. But he needed a better understanding of zinc ores, and for that he turned to Smithson's article.

As part of the preparations for his enterprise, Dony is known to have provided support for several important scientific studies, one of which appears to have been the translation of Smithson's calamine article, which appeared in the *Journal des Mines* in 1810. This was the leading French journal for mineralogy, and the man who edited Smithson's article was Pierre Berthier, a twenty-eight-year-old mining engineer who would go on to have a distinguished career. Berthier's version of Smithson's article begins with an unusual statement. In the first footnote, on the first page, he bluntly states, "I have significantly abbreviated the Memoir of Smithson and I have changed its order; but I have only suppressed that which does not have a direct connection with the principal object, the knowledge of calamines, and I have scrupulously kept the sense of the text."

To be more specific, what he suppressed was every reference to Smithson's theory, either direct or indirect. Nearly half of Smithson's "Observations" section—which he would have considered the real substance of his work—is simply missing. While Berthier claimed to do this in order to concentrate on "the knowledge of calamines," he felt free to expound at length on why "the hypothesis of M. Smithson is refuted by the results of several analyses made with great care."

Berthier added some interesting material as well. He replicated all of Smithson's tests, with a much greater number of samples, and confirmed essentially every step of Smithson's chemical analysis. He also told a story about how Smithson's identification of zinc carbonate had been received in Paris: "Before the publication of the work of M. Smithson, mineralogists mixed up all the calamines into one single species, the nature of which they were not able to agree on. It is all the more singular that, in general, they decided to reject the carbonate of zinc, even though this substance is precisely the most common. Out of ten samples of calamine, taken by chance from a [mineral] collection, there were at least seven that belonged to carbonate." The evidence had been right in front of them, but only Smithson had seen it.

The *Journal des Mines* article was the only version of Smithson's article that most French savants would ever see, and this meant that many of them were probably unaware of his theoretical work. Smithson could hardly have been happy about this, but it must be said that once it was freed from his theory (which by 1810 was getting dated), this new version of "Calamines" became much more accessible. It also speaks to how much Smithson put into it that, even in this form, the article was widely recognized in France as a fundamental scientific work. "A Chemical Analysis of Some Calamines" became the starting point for all subsequent French studies of the zinc ores.

As for the production of zinc, the Napoleonic Wars ended just as the mines at La Calamine began producing high-quality zinc in significant amounts. This came too late to be of use to Napoleon, but the new metal promised to be so important that in the subsequent peace negotiations it was thought to be imprudent to let the mines fall under the control of any one nation. So in 1815 the area was given special status as "the Neutral Territory of Moresnet," which meant that it effectively became an independent country, albeit a very small one. Moresnet had a circumference of only about nine miles, and it shared borders with Belgium, Prussia, and the Netherlands. Moresnet supplied Europe with a seemingly inexhaustible supply of the metal throughout most of the nineteenth century.

By the early twentieth century other sources of zinc had been developed, and Europe no longer needed a separate country dedicated to mining it. In the reorganization that followed the end of World War I, Moresnet was abolished and the territory became part of Belgium. Today the area is mostly farmland, and only a few traces remain of the vast industrial enterprise that once existed.

Smithson's calamines article was ultimately most important to mineralogy, where Smithson's name soon came to be identified with the zinc ores. Martin Heinrich Klaproth's influential *Dictionnaire de chimie* (1811) carried a long discussion of them, with prominent mention of Smithson. In 1814 Thomas Thomson, the respected editor of *Annals of Philosophy*, credited Smithson with discovering three of the four known ores of zinc. In Europe, much more so than in England, Smithson was also credited for his particular interest in hydrates.

In the end, perhaps the most reasoned assessment of Smithson's "Calamines" grew out of the correspondence between Smithson and the great Swedish chemist Jacob Berzelius, part of an ongoing discussion they maintained about Smithson's ideas. In their few surviving letters we find Smithson giving details of how he had made the calamine analysis and even sending Berzelius his last sample of the same zinc carbonate crystals he used in his original tests. This was in 1818, and in an article published the following year, Berzelius praised Smithson's "excellent work, published fifteen years ago, upon the composition of the different ores of zinc, which are named calamines." Then he kindly but firmly pointed out that Smithson had been mistaken about the composition of the Bleyberg calamine: "Mr Smithson himself considered this substance as a chemical combination of the carbonate and the hydrate of zinc; and the inaccuracy of the proportions which he assigns to these ingredients must have proceeded from his unacquaintance with those laws of combination which have since been discovered."

"Calamines" was only Smithson's second published article, and it came twelve years after the first. It was an ambitious project, and he clearly put an enormous amount of work into it. It must have been a real disappointment for him that, because of the war, the French remained largely unaware of it, and especially about what he would have considered to be its most important parts. The discussion of his chemical theory that he clearly wanted to initiate never happened. Yet "Calamines" is generally considered today to be his most important paper, and it is the reason that the mineral zinc carbonate was later named "smithsonite" in his honor. But in 1803 all these things were still in the future. With "Calamines" finally published, Smithson went to Paris for a vacation—and his life took another unexpected turn.

———————|———————

5.

MINIUM

SMITHSON'S CALAMINES ARTICLE was published in the spring of 1803 during the Peace of Amiens, the short-lived truce between England and France. The Peace had started about a year earlier, and many of his friends had joined the ensuing crush of British tourists that headed across the Channel to Paris. Smithson seems to have stayed behind until his article came out, and it was late April 1803 before he was finally able to join them. May 16 saw the resumption of hostilities between England and France, however, and it was announced that on May 22 all English males still in France would be declared prisoners of war. As Smithson put it, "I was so unfortunate as to get into France just at the time that it became necessary to leave it."

He narrowly escaped. Either unable or unwilling to return to England, he headed north and left French territory only two days before the arrests were set to begin. France's borders had swelled since the Revolution, and he needed to get to what is now the eastern part of the Netherlands to be safe. From there he traveled east into what is now Germany.

By late 1803 Smithson had made his way to Cassell (Kassel). The city had a small English community and many scientifically minded residents, and there Smithson tried to make the best of his situation. He was still able to send and receive letters, and he seems to have had access to his bankers, so money was not a problem; but the mail service was less than reliable, and the few scientific journals that reached him took a long time to arrive. He stayed in Cassell throughout much of 1804, unable to return to England because the route across the Channel was now closed. In the spring of 1805 he headed south toward the Alps, visiting Basel, Neuchâtel, and Zurich, and by late July was at Karlsruhe. By November, the movements of French forces had pushed him back to Cassell.

By 1806, Smithson had been away from England for nearly three years. Even though he was relatively safe as long as he stayed in the German territories, Smithson was cut off from his friends and colleagues in a way that is hard to imagine today. Before the war between England and France began, communication within the international scientific community depended partly on personal contacts but mostly on personal correspondence. While not inexpensive, the mail service was reliable, and most Enlightenment savants were prolific letter writers. Their exchanges would often continue for years, and it is no exaggeration to say that the friendships and collaborations that developed out of them were the glue that held the international scientific community together.

The wars played havoc with the postal services, and by 1806 whatever correspondence Smithson had with his friends back in England was mostly accomplished by avoiding the post entirely and entrusting his letters to other travelers, who hand-carried them back across the Channel. Inconvenient as it was, the system worked, although judging from the surviving correspondence, it worked far better for letters coming *from* Germany than for those going to it. Letters from Smithson generally arrived in London just a week or two after being written, but an important issue of the *Philosophical Transactions* took two years to make the opposite journey from London to Cassell.

Smithson was rapidly losing touch with the London community, and in letters to his friends he talked about planning to return to England soon. He was waiting for the right opportunity, but before that came, he made the discovery that led to his third article, "Account of a Discovery of Native Minium" (1806).

In contrast to his first two publications, which were long, closely reasoned, and thoroughly researched, this article was submitted in the form of a quickly written, two-page letter. Letters from traveling members were a tradition at the Royal Society, providing a convenient way to report new observations and discoveries. All such letters went directly to Joseph Banks, the Royal Society's president, who determined their fate. Most were treated like regular correspondence, but if they were interesting, he could have them read at a society meeting; if they were exceptionally interesting, they might even be published. Banks made sure that Smithson's letter was read to the society, and then he promptly published it.

What Smithson found was a sample of lead ore from a mine in Germany. The ore, a familiar combination of lead and sulfur called galena (PbS), had been deposited on a layer of limestone, coating it with distinctive, silvery-gray crystals. This was not unusual. What *was* unusual was that some of the crystals were themselves covered with a thin layer of a reddish material. Smithson initially assumed the material to be in a "pulverulent state," like a dusting of fine powder. But when he examined it with a lens, he found that the reddish material had a "flaky and crystalline texture," indicating that it had formed in place. Perhaps even more important, he recognized the particular color of the material as being that of minium.

Better known today as "red lead," minium is the Latin name for a pigment whose use dates back to antiquity. Chemically, it is an oxide of lead (Pb_3O_4), readily identified by its distinctive color, which Smithson accurately described as "a vivid red with a cast of yellow." It was still a useful material in Smithson's time and continued to be manufactured in significant amounts. Minium also enjoyed the unusual distinction of being one of the few known substances that did not seem to occur in nature. Thus, with the exception of the material on his sample, all the minium known to exist had been made by humans, a fact that Smithson referred to when he described the color of his sample as being "like that of factitious [manufactured] minium."

Like many industrial processes, the manufacture of minium had been studied by chemists and would have been familiar to most of Smithson's readers. Basically, it was produced by two controlled applications of heat. The first started with powdered, metallic lead, which was heated in a furnace until, at around 890°C, it combined with oxygen to form litharge (PbO). This chemical change was identified by the lead turning from a silvery-gray color to a canary yellow. After cooling, this material was washed and crushed into a fine powder and then heated a second time. But this time it was first spread out on the hearth of a coal-fired furnace, to put it into contact with the air. Then it was continuously heated (at about 450°C) for at least forty-eight hours. This slowly converted it to minium, a transformation confirmed by its turning a distinctive reddish-yellow color.

Smithson used this manufacturing process—in reverse—to identify his sample. Starting with a small amount of the suspected minium, his first step was to heat it with the blowpipe. Blowing gently and holding the sample well away from the tip of the flame, he reported that in this moderate heat the sample "assumes a darker colour, but on cooling it returns to its original red." This was a common blowpipe test, and a strong indication that the sample was a metallic oxide.

The next step was to try to convert it to litharge. Smithson reported using "a stronger heat," which probably means that he was now blowing harder through the blowpipe and holding the sample a little closer to the tip of the flame. With practice it is possible to control the flame so that the entire sample melts and then almost immediately cools, and in that instant it is transformed from the reddish-yellow minium oxide into the canary-yellow litharge oxide. What happened here was that Smithson heated the sample to the point that some of the oxygen escaped; when it cooled, the remaining lead and oxygen combined in a simpler form: Pb_2O_3 (red lead) became PbO (litharge).

The last step was to reduce the litharge to pure metallic lead, and Smithson accomplished this by blowing even harder through the blowpipe and bringing the tip of the blue flame into direct contact with the sample. In this tiny, brutal environment, the sample melted and the yellow color completely disappeared. As the sample cooled, it formed a small silvery-gray ball, and Smithson reported simply that "on the charcoal it reduces to lead."

It was a convincing analysis, and if the ball was truly lead, it reliably identified the reddish material as minium. Smithson accordingly now examined the ball and he reported that it dissolved quickly in nitric acid, but that it turned the clear liquid a "coffee colour." He found that adding "a little sugar" to the solution made it clear again. Smithson did not bother to explain that this was a known test for lead. He expected his readers to know that the sugar had reacted with the solution to produce oxalic acid, which in turn had joined with the lead in the sample to precipitate lead oxalate, leaving the nitric acid pure and just as strong as it had started out being. This test thus confirmed that the ball was composed of lead—as long as the nitric acid was still potent.

Smithson's final step was to demonstrate the strength of the acid, which he did by a clever use of "Aqua Regia," Latin for "royal water," a highly corrosive, fuming liquid made by combining nitric acid with hydrochloric acid. It cannot be stored, but when mixed fresh with strong ingredients, it is one of the few reagents able to dissolve gold. Smithson combined his nitric acid with hydrochloric acid to make aqua regia, and then successfully dissolved a slip of gold leaf in it. This simple act proved the potency of the nitric acid, and thus completed the chain of reason and experiment that identified the reddish material he had found to be minium.

Smithson's analysis was well conceived, particularly given the small size of his sample, and there seemed to be no question that it was correct. But the discovery of natural minium raised the question of how it had formed, and Smithson had an answer for this as well.

It took high temperatures to manufacture minium, but Smithson's samples showed no evidence of having been burned or strongly heated, so the question of how they could have formed was of particular interest. As it happened, one of them provided him with a clue. The sample contained a cluster of what appeared to be solid minium crystals, but when he broke one open he found that its center was galena, the silvery-gray ore of lead. On examination with a lens, it appeared that the entire crystal had originally been galena, but that the outer surfaces had somehow been converted into minium. This discovery led Smithson to suggest a novel method by which it could have formed. "This native minium seems to be produced by the decay of a galena," he wrote, "which I suspect to be itself a secondary production from the metallization of white carbonate of lead by hepatic gas."

What Smithson suggested was a process in which an existing lead ore, "white carbonate of lead" ($PbCO_3$), had been transformed into the lead ore, galena, by contact with "hepatic gas"—a foul-smelling combination of sulfur and hydrogen that occurs naturally in mines. And, without going into details, he suggested that the outer surfaces of this galena had subsequently decayed to form minium. It was an ingenious suggestion, and it seemed to offer the best explanation for the curious nature of the minium deposits, but for the most part it was ignored by his readers. Most naturalists of the time still preferred to confine their speculation to the two accepted agents of mineral formation: solvents and heat.

Smithson's discovery of natural minium was quickly and widely acknowledged. His article was reprinted in at least three other scientific journals and reported in most of the rest. Given his reputation, there was no challenge to his analysis, and apart from the tepid reaction to his speculations about hepatic gas, the only real question about his discovery was why it had not been made sooner. Lead had been actively mined for more than two thousand years; why had Smithson been the first to notice any minium on it? This sentiment found expression in the long account of his article that appeared in the *Critical Review*: "native minium, confessedly a rare mineral, is asserted to arise from the spontaneous decomposition of sulphuret of lead [galena], though the latter body existing in vast abundance, and exposed to every cause of destruction, had never before been observed to exhibit a similar phenomenon. We cannot subscribe to such an opinion."

It was a fair point, and part of the explanation is simply that minium deposits are both rare and extremely fragile. In some cases the thin layer of minium could be brushed off by contact with a miner's glove. But the fact is that once Smithson reported finding it in Germany, minium began turning up in the mines of other countries as well. The rarity and philosophical interest of these samples made them prized by collectors, and once miners knew what to look for and what they were worth, the delicate specimens began to be found. In 1809, just three years after Smithson's article, a color image of a minium specimen from an English mine appeared in James Sowerby's authoritative *British Mineralogy*.

The discovery of natural minium was a significant find, and it certainly added to Smithson's reputation, but had it not been for the discovery of a clue about how it formed—had there not been some "philosophical" side to the discovery—it is unlikely that he would have written about it.

In the holdings of the British Library there is a letter from Smithson written in the early 1790s to his friend Charles Greville, the mineral collector. In it, he repeatedly brings up the topic of minium, or "red lead." In the body of the letter he writes: "I particularly wish to know whether you have in your Collection, or have ever seen, any like of Cumberland red ore of lead adhearing to its stony matrix, or which bore any other indubitable marks of its being a natural production. Also whether you have this substance in regular crystals, such is said to have been found—some experiments which I have lately made on this substance, & which show it to be perfectly analogous to a calx of lead produced in our furnaces, lead me to trouble you with these inquiries." And then again in a postscript: "I wish you would cast an eye upon my Blocks[?] of Cumberland red lead—& see whether you can find any certain traces of the fingers of nature upon them, for my part I cannot, & begin to be apprehensive that they will turn-out to be products of art [i.e., manufactured]." It seems clear from this letter that Smithson had come across some specimens of natural minium from one of the lead mines in Cumberland, England, and was in the process of making a study of them. But this letter was written at least fifteen years before Smithson announced his discovery of minium in Germany.

The missing part of this story is that in the early part of his career, Smithson appears to have undertaken a series of detailed studies of various mineral groups, including the ores of lead. These studies were made for his own use; none were published and none survive. They seem, however, to have given him an extraordinary familiarity with the mineral kingdom that informed much of his later work. It appears that he made the minium discovery as part of his study of lead oxides, but chose not to publish it until he could identify the method by which it had formed, which he finally did in Germany fifteen years later.

One last aspect of Smithson's minium letter that deserves mention is that before it was published, a short account of it appeared in *The Philosophical Magazine*. Written by an unnamed reviewer who was apparently present at the Royal Society meeting where it was read, the account was favorable but contained a number of errors—the most significant being that it identified the letter's author as Smithson Tennant.

It was an understandable mistake. Like James Smithson, Smithson Tennant was an accomplished chemist and a member of the Royal Society. Only four years apart in age, the two men were friends and had many friends in common, so they could have easily been confused. A correction promptly appeared in the next issue, which should have been the end of the matter, but having appeared in print the mistake took on a life of its own. A few months later it was repeated in *The Scots Magazine*, in Edinburgh, and in 1809 it was enshrined in *British Mineralogy*, which cited Smithson's article correctly but identified Smithson Tennant as its author.

This seems to be the first example of James Smithson and Smithson Tennant's being confused for each other, but it would not be the last. Foreign journals, particularly the German journals, had a difficult time distinguishing between the two chemists and frequently misattributed their work. While there is no evidence that this was of any concern to Smithson, it later became an abundant source of confusion for historians trying to understand his work.

In the end, Smithson's minium article was a great success. It was an important and surprising discovery, and it kept him from being forgotten back in England during his long absence. But it did not relieve his isolation or the pressure he must have felt as Napoleon began to bring the German states under French control. In November 1806 the French army took Kassel, and in 1807, Smithson finally decided to try to return to England. But he had waited too long and instead was arrested—caught up in "the hurricane of war," as he put it. By 1808 he was in Hamburg, imprisoned by the French invaders, in deteriorating health, and with no prospect of release.

In September, having exhausted all his other options, Smithson wrote another letter to Joseph Banks, this time asking for his help. This was a much different letter from his minium report, and Smithson began with a surprisingly emotional description of his situation. "I am really here in a most untoward situation," he wrote, "in fact an utter stranger to every body, deserted by those on whom I had depended, not perhaps to say worse, & vibrating between existence & the tomb."

Having stated this, he abruptly changed his tone. What was important to men like him and Banks was the pursuit of science, and the real danger was that Smithson's scientific work might be lost. The minium article had been just one of his projects, and to make that point Smithson provided Banks with an inventory of what he had been working on:

I wrote a paper at Tonningen a little before being taken ill, which I have promised in a letter to M. Greville, to send him the moment I can hunt for it. I have several others in some forwardness, I have also collected a considerable mass of detached notes & observations, & I have a certain number of new substances. I have besides with me many of the papers of two more considerable works on which I have been long engaged. I wish much to get to England to arrange & finish them, as I should be sorry that they were all lost by my death after all the pains & time they have cost me.

On receipt of Smithson's letter, Banks quickly took action. He held no official governmental position, but as the longtime president of the Royal Society and an astonishingly active correspondent, Banks was one of the best-connected men in Europe. On several occasions he had gone out of his way to assist French savants caught up in the events of war, and now he wrote a letter to Jean-Baptiste Delambre, his counterpart at the Académie des Sciences in Paris, asking for help with Smithson. It took time, but his letter was finally read to the members on March 20, 1809, and forwarded, at their request, to the French government. It took more time for the request to work its way through the French bureaucracy, but three months later Delambre was able to report back to the members that Smithson's case had been reviewed and Napoleon had ordered his immediate release. It is important to understand that this kind of release was far from a common event. Despite the respect that both governments proclaimed for the pursuit of science, the total number of savants receiving special treatment during the course of the Napoleonic Wars hardly exceeded a handful on either side.

As promised, Smithson's new passport was delivered to him in Hamburg, and by midsummer of 1809 he was free to return to England. But Smithson was no longer in a hurry. Instead he headed southeast to Berlin. The purpose of the trip is a bit puzzling. Perhaps he wanted to visit his old friend, the chemist Martin Klaproth, or wanted to make a point to any French agents who might be following him. For whatever reason, the year was nearly over by the time Smithson finally returned to London. He had been gone for more than six years.

———————————|———————————

6.

THE SULPHURET
FROM HUEL BOYS

THE YEAR 1809 WAS NEARLY over by the time James Smithson finally returned to London. He had been gone for more than six years, and, not surprisingly, it took him some time to pick up the pieces of his life. He needed to put his affairs in order, find and furnish a home, and digest the reaction to his latest article, which he had sent from Germany more than two years earlier. Although again elected to the Royal Society's governing council shortly after returning, he did not attend any meetings until the following June.

Titled "On the Compound Sulphuret from Huel Boys, and an Account of Its Crystals," this article was partly a defense of the chemical theory he had put forward in his calamines paper and partly an extension of it. To fully understand it, the story needs to be seen in the context of London at the turn of the nineteenth century and the early years of the Royal Institution.

The Royal Institution, founded in 1799, brought a new energy to the London scientific scene. This was particularly true after 1801, when chemists Humphry Davy and Thomas Young received appointments. Both men would soon play outsized roles in British science, but in the early years of the nineteenth century their first impact was to make the Institution a hub for chemistry, particularly philosophical chemistry, and in turn one of the sites where the new atomic theory would be developed.

Thomas Young, the institution's new professor of natural philosophy and chemistry, was one of a group of speculative English chemists that historians have come to call "the London atomists." The "atomists" were an informal group centered around Bryan Higgins, who from 1774 to 1797 ran a private chemistry school out of his home in Soho. Chemistry was a popular subject at this time, and in addition to introductory classes for young men, Higgins also offered an advanced course "composed for the use of the noblemen and gentlemen." Presented once a year and held in the evening, the course covered the full range of late-eighteenth-century philosophical chemistry. Higgins would begin each class with a short lecture on a topic of theoretical interest, followed by demonstrations of the relevant chemical phenomena, and then an open discussion. This novel approach to teaching chemistry attracted considerable interest. Benjamin Franklin, who was living in London at the time, attended some of Higgins's early lectures, as did the chemist Joseph Priestley, who enthusiastically recommended them to his friends. Thomas Young, who had attended Higgins's chemistry school as a boy, was later employed by Higgins as an "assistant experimenter" in some of the advanced courses.

Over time, Higgins's school generated a small community of English natural philosophers with an advanced understanding of chemical theory—the "Atomists." Even if they were not in complete agreement with all of Higgins's opinions about atoms and the ways in which materials combined, they were committed to the further exploration of these topics. When Higgins's school closed in 1797, many in the group looked for a way to keep the discussion going. Their goal was to merge with an existing society, and in the early 1800s several of them, including Young, found a home in the newly founded Royal Institution.

Smithson also had connections to the atomists. It is likely he attended Higgins's school prior to enrolling at Oxford, and while at Oxford he knew at least two important members of the group. There were also connections between Higgins's school and the Coffee House Philosophical Society, which Smithson had attended, and he and Young were both members of the Royal Society. But the two had the most opportunities to interact at the Royal Institution. As one of the institution's founders, Smithson would have been welcome at any of Young's talks, and he could easily have arranged to chat with him privately. Whether or not he did is unknown, but it is worth noting that Smithson's clearest statement of his ideas about chemical bonding was in his calamine article, which he seems to have started writing in 1802, shortly after Young joined the Royal Institution.

In the early years of the nineteenth century, London and particularly the Royal Institution were at the center of a movement to find a new understanding of matter and the chemical bond. Smithson was located both physically and intellectually within that movement. It was in this context that his calamine and sulphuret articles were written and should be understood. This can be seen in a review by Thomas Young of an article by the French chemist Joseph Proust, who had been experimenting with metals and the ways in which they combined with sulfur. One of the properties of sulfur is that it joins readily with metals and sometimes forms exotic minerals—called "sulphurets"—in which it combines with two or even three different metals. Although Smithson's name did not appear in Proust's paper, it seemed clear to Young that Proust had read it, and he accordingly began his review by stating that Proust's article "may be regarded as an extension of his [Smithson's] ideas on the composition of ores to the metals mineralized by Sulphur."

Proust had been attempting to chemically extract the "immediate principles" of these sulphurets—the individual binary compounds that, in Smithson's view, formed first and then combined with each other to form the final mineral. In retrospect, it was an ill-considered plan. The kind of chemical analysis Proust applied to these minerals, basically breaking them down with acids and heat, did not easily distinguish between "principles" that already existed in the minerals and new "principles" produced by the analysis itself. Inevitably the results of his experiments were so ambiguous that he returned to an old idea: that metals sometimes "dissolved" in minerals. His conclusion was that "at least in the more compound minerals, one of them serves as a solvent which dissolves the others in an infinite number of proportions below the maximum, as in the case of salts dissolved in water." Proust did not argue against binary combination per se, but rather presented these compound combinations as special cases, in which the rule of strictly binary combination did not apply.

This question of how the metals combined with the sulfur in the sulphuret was on the cutting edge of philosophical chemistry. Proust's conclusion that the metals were "dissolved" harkened back to some of the atomist arguments, and it stood in direct contrast to the theory Smithson proposed. Smithson clearly thought of chemical combination as being "atom to atom" (to use William Wollaston's later phrase, in which "atom" roughly corresponds to our idea of an "ion"), a view that is much closer to the modern understanding. It was an important conceptual leap, and in his review of Proust's article, Young credited Smithson for "being the first that made any decisive observations on the subject."

Interest in these kinds of philosophical questions was strong at the time, and Young used his review to comment on some of Smithson's other assumptions, informing his readers that, along with Proust, he agreed "perfectly" with Smithson's views about binary combination (albeit with some possible cases of "ternary," or three-way, combination), but that he completely rejected Smithson's idea of substances always combining in simple ratios, which he called "vulgar fractions." Throughout his review, Young's familiarity with Smithson's ideas and the fact that he took them seriously is obvious.

Young wasn't the only one at the Royal Institution interested in philosophical chemistry; his colleague Humphry Davy organized public lectures about the latest theoretical developments. In the summer of 1803 William Allen, who had attended Higgins's school, presented a series of chemistry lectures that included a "summary of the particles of matter." John Dalton was also invited to the institution, and beginning in December 1803, he gave a remarkable series of nineteen lectures on the status of his investigations and the state of chemical knowledge. Just five years later Dalton would bring all the pieces together in *A New System of Chemical Philosophy*, the groundbreaking work on atomic theory that provided the foundation for modern chemistry. Given his interests, one would expect Smithson to have attended these talks, but he was stranded on the continent, safe in Germany but increasingly out of touch with the fast-moving developments back in England.

During his time in Germany, Smithson's only source of information about developments in England came from scientific journals, which often arrived very late. Moreover, he had no way to know which articles were considered important. He doesn't appear to have known of Proust's article, or Young's review of it. Toward the end of 1806 he finally received a copy of the 1804 *Philosophical Transactions*, where his attention was drawn to a long paper by French mineralogist Comte de Bournon.

Bournon was a former French military officer who found it prudent to move to England after the French Revolution. Once there, he turned to mineralogy as his occupation. He was knowledgeable, with the eye of a connoisseur, and he supported himself by organizing and expanding the private mineral collections of London gentlemen—an occupation that speaks to both the size and number of such collections at this time.

His article in the *Transactions* carried the long title of "Description of a Triple Sulphuret, of Lead, Antimony, and Copper, from Cornwall: With Some Observations upon the Various Modes of Attraction Which Influence the Formation of Mineral Substances, and upon the Different Kinds of Sulphuret of Copper." It described his investigation of an unusual and little-known mineral recently found in the "Huel Boys" mine in Cornwall. Chemically the mineral was another sulphuret, a mineral containing sulfur and, in this case, the metals antimony, copper, and lead. Bournon was a mineralogist and a crystallographer, but he was no chemist. He persuaded Charles Hatchett, a very good chemist, to investigate it. Hatchett's analysis followed Bournon's article in the same issue of the *Transactions*.

Hatchett described the interesting series of effects that could be produced by heating this mineral with a blowpipe. If you melted one of its small, dark crystals on a piece of charcoal and then let it cool, it formed a solid, dull-gray lump. But if you kept heating it, the mineral would begin to bubble and release white fumes (antimony) as well as the distinctive "rotten egg" smell of sulfur. Continued heating caused the sample to shrink in size, eventually forming a mass that, when cooled, had a crust of "sulphuret of lead" which, when broken open, was found to contain a tiny ball of almost pure copper.

This experiment provided an intriguing analogy with the process of dissolving. Specifically, the analogy was with a bowl of water in which different chemical salts have been dissolved. As the water slowly evaporates, each salt is successively deposited out of solution as its particular saturation point is reached. This, it could be argued, was analogous to what happened in Hatchett's blowpipe test. As the sample was heated, and as the sulfur started to boil away, the antimony was the first to come out of "solution," which it did as white fumes. Next was the copper, which precipitated out of solution and formed a small ball, leaving a dark material that consisted of lead and sulfur, still in "solution."

From this and other tests, Hatchett concluded that the three metals in the mineral "exist in the ore, in, or nearly in, the metallic state, combined with sulphur, so as to form a triple sulphuret." In essence, he thought the three metals were "dissolved" in the sulfur, and if this was true, it seemed that an exception had been found to the rule of binary combination. Hatchett did not specifically mention Smithson in his article, but his analysis certainly seemed to call Smithson's theory into question.

Bournon incorporated Hatchett's conclusion into his own paper, adding it to his description of the mineral's physical characteristics. But as a crystallographer, his main interest was to determine the "primitive form" of the crystal, which he reported to be "a rectangular tetrahedral prism, which has its terminal faces perpendicular to its axis." Most of the rest of Bournon's article was taken up with the elaboration of an unlikely theory of mineralogical (as opposed to chemical) combination. It was a curious piece of work that one historian called "a mineralogical affinity theory," and it generated little interest. Smithson did not know this, though, and when he read Bournon's article he must have seen it as a challenge to his own theory.

Smithson's reply was the aforementioned article "On the Composition of the Compound Sulphuret from Huel Boys, and an Account of Its Crystals," which he hurried to complete and probably sent off to England in the summer of 1807. The first thing he addressed was Hatchett's conclusion that the Cornwall sulphuret was not binary. Using data from Hatchett's own analysis, Smithson put the sulphuret through the same theory-based analysis that he had employed in the calamine article, and with similar results. He was able to show how the "triple sulphuret" could have formed by a series of binary combinations.

In the second part of his article, Smithson addressed Bournon's statements about the mineral's crystallography. Although he was confined to Germany, he had somehow gotten a few small crystals of the sulphuret, and his measurements of their angles allowed him to correct Bournon's. He declared that the "primitive form" was a cube rather than a rectangular tetrahedron, as Bournon had claimed, and he criticized the French mineralogist's other measurements as "inconsistent."

He saved his harshest criticism for Bournon's theory-based crystal drawings. Smithson wrote: "Of the seventeen figures which have been given, as of the crystals of this compound sulphuret . . . [the] great part are acknowledged to have no existence, nor are indeed any of them consistent with nature." He criticized Bournon's inclusion of so many theoretical forms (whether correct or not) as unnecessary and "not only superfluous, but most truly puerile."

While his criticisms seem to have been justified, Smithson's treatment of Bournon, whom he had met, was uncharacteristically unkind. He may have simply been expressing his frustration at being away from London and unable to address this challenge directly, but there may have been another factor. Smithson was an adherent of Haüy's crystallography, which became a key part of his science. However, Bournon had studied with Haüy's old rival, the crystallographer Rome de l'Isle, and he still followed de l'Isle's crystallographic system. De l'Isle died in 1790, long before Haüy published his landmark *Traité de minéralogie* (1801), the most complete expression of his theory of crystals. Haüy's theory was widely admired and seen as supplanting all others in the science. By the time that Smithson wrote his sulphuret article, he may have felt that any crystallographer still clinging to de l'Isle's system could not be taken seriously.

In any case, Smithson's poor choice of words would soon cause him problems. His article arrived in London at the house of William Wollaston, a secretary of the Royal Society and one of the people who decided which articles would be published in the Society's journal. It arrived while Wollaston was entertaining a friend—none other than the Comte de Bournon. As Bournon later wrote, "Chance made me acquainted with the criticism of Mr. Smithson, at the time of its being delivered to the secretary of the Royal Society, Dr. Wollaston, at whose house I then happened to be. He gave me permission to look it over. Its nature surprised me."

Bournon's calm indifference did not last long, because he seems to have unsuccessfully attempted to suppress the publication of Smithson's article. The

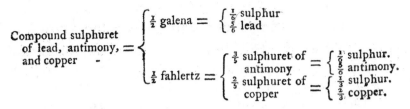

Smithson's binary representation of how he thought the compound sulphuret from Huel Boys was composed. He called the mineral a "compound sulphuret" in preference to Hatchett's "triple sulphuret"—a term he would not use. The mineral's "immediate principals" and their "simple" proportions are listed on the right. Courtesy of the Smithsonian Libraries.

article was read to the Society on January 28, 1808, and published shortly thereafter. Bournon submitted a response to Smithson's article, which, after his repeated requests, was eventually read to the Society but never published. Eventually he published a long, rambling article in the *Journal of Natural Philosophy* that began: "This memoir was written chiefly as an answer to that printed in the first part of the *Philosophical Transactions* for 1808, in which Mr. Smithson its author, criticizes with as little justice as decency a former memoir of mine." Bournon was well connected and active in several of London's new scientific societies, and he now saw Smithson as his enemy.

Apart from Bournon, not many people commented on Smithson's "sulphuret" paper, but those that did mostly remarked on Smithson's lack of respect for his distinguished colleague. The unidentified reviewer for *The Eclectic Review*, for example, found no fault in Smithson's correction of Bournon's measurement of the mineral's "primitive form," but he complained, "In demonstrating the truth of this assertion, Mr. Smithson pays little respect to the well known abilities of Count Bournon, with whose skill in the field of mineralogical science the world is well acquainted. The tenor of this part of Mr. Smithson's paper is certainly not becoming a philosopher of his general respectability." Thomas Young likely wrote the review of Smithson's article for the *Retrospect* and commented, "The acknowledged crystallographic abilities of Count Bournon should certainly have led Mr. Smithson to have spoken of his description of the crystals of this triple sulphuret with more respect."

The critics also disapproved of Smithson's use of the common term "powder of Algoroth" for one of his chemicals (antimony oxychloride, $SbOCl$). This was the name apothecaries used when they sold it as a medical purging agent, and this "barbarism," the critics argued, was not appropriate for scientific use. A more substantial objection was put forward by Bournon himself, who gleefully reported finding two errors. Smithson had used the data from Hatchett's analysis in his own article, but to complete it he had also needed to know the "combining weights" of each of the three metals with sulfur—which Hatchett had not provided. Smithson, who was probably in Kassel at the time, attempted to make these determinations himself, but he was hampered by the unavailability of certain materials. He had been forced to estimate the combining weight of copper, and his estimate turned out to be incorrect. Bournon reported that Smithson's determination of the combining weight of antimony had also been wrong. These uncharacteristic errors must have been embarrassing for Smithson, especially coming from Bournon, but they did not necessarily refute his arguments or his theory. Thomas Young wrote that he considered Smithson's binary explanation of the sulphuret as "ingenious and probable."

But Smithson's article had bigger problems. Because of his isolation in Germany, he was unaware of developments back in England, which included the chemical theory being developed by John Dalton. The first printed exposition of Dalton's atomic theory appeared in 1807, published in Thomson's widely read *System of Chemistry*, and Thomson had become an early supporter. Dalton's own version of it had been published the following year and was on everyone's mind when Smithson's sulphuret article was published. Although he had written it to refute Bournon's theory, Smithson's article was evaluated on the basis of how it compared to Dalton's.

For example, Young wrote that, although it seemed that all substances combined chemically in simple proportions, as Smithson suggested, "Yet we cannot but think that Mr. Smithson's view of the cause of this effect is far inferior to that of Mr. Dalton." Smithson had originally assumed that substances combined in specific proportions, by weight, but one of Dalton's great contributions had been the concept of "atomic weight"—that the atoms of each element have a unique and specific weight. This fundamental concept promised to become a powerful tool once the atomic weight of each element was determined. Young observed, "Although each hypothesis [of Smithson and Dalton] will equally serve to explain the phenomena of nature, and to correct the errors occurring in the manipulations of practical chemists; that of Dalton is probably the real one." This seems to have been the general conclusion.

In addition to bad timing, Smithson's article also suffered from bad placement. Immediately following it in the same issue of the *Philosophical Transactions* were two papers that would each become landmarks in the history of chemistry. The first was Thomas Thomson's "On Oxalic Acid." Thomson was perhaps the most active supporter of Dalton's ideas, and this paper gave the theory its first predictive success. The second article was William Wollaston's "On Super-Acid and Sub-Acid

Salts," which he had rushed to complete in time to include it in the same issue. Like Thomson, Wollaston was an enthusiastic supporter of Dalton's theory, and in this article he experimentally established the law of multiple proportions. Taken together these two papers provided powerful experimental confirmation of Dalton's theory.

Smithson's sulphuret article was completely overshadowed by these papers, and it is no wonder that one historian called it a "work of little long-term importance in a forgotten paper." Indeed, it would be hard to argue that this paper was in any way a success. Smithson never responded to the errors that Bournon found in his calculations, and the sulphuret article would be his last direct statement about theoretical chemistry.

Determining Smithson's precise role in the development of atomic theory has proved challenging, in part because Dalton was reluctant to admit the influence of other researchers. But many of Smithson's contemporaries had thought he made a contribution, and historians of chemistry have noted the suggestive timing of Smithson's calamines article, published early in 1803, just as Dalton began his important experiments on the combining proportions of gases.

In 1812, in a long, critical review of Humphry Davy's *Elements of Chemical Philosophy*, the unnamed reviewer spoke of recent advances in chemistry. He corrected Davy's failure to mention several of the figures who had played important roles in the establishment and extension of the law of definite proportions, writing that "for these facts the science is principally indebted after Mr. Higgins, to Dalton, Gay Lussac, Smithson, and Wollaston."

More recently, it has been suggested that Smithson's successful use of simple ratios, his "emphasis on the small numerical fractions expressing chemical composition," as historian of science Robert Siegfried put it, may have influenced Dalton to include numerical simplicity in his theory, particularly in his conception of equivalent weights. And Smithson may have contributed to atomic theory in an even more fundamental way. His calamines article described chemical combination in terms of its being "atom to atom," and this was a critical distinction that no less an authority than Thomas Young credited Smithson with being the first to make. John Dalton later assumed atom-to-atom combination as part of his atomic theory.

For his part, Smithson appears not to have accepted some of Dalton's ideas. In particular, he seems to have continued to argue the case for strictly binary chemical combination—although not in print. In a complex chemical analysis that he published in 1813, he followed the same protocol of using "combining weights" that he used in his calamines article. And although he did not call attention to it, all the materials he described in that analysis were binary. Six years after that, in a letter to the great Swedish chemist Berzelius in which he talked about his work on calamines, Smithson revealed that he still considered all chemical combination to be binary when "properly considered."

Smithson's reluctance to express his objections to Dalton's theory is puzzling, because he was clearly thinking about it, but he would never write about chemical theory again. This silence is hard to explain without reference to the deep frustration Smithson must have felt. The development of atomic theory had been a once-in-a-lifetime opportunity for him to be part of a revolution in chemistry and, had he been in England, Smithson might have been an important voice in the scientific conversation that produced it. But his ill-timed trip to France and his long isolation had robbed him of the opportunity.

———————+———————

7.

ON THE COMPOSITION
OF ZEOLITE

AROUND THE MIDDLE of the eighteenth century, naturalists began to study the distinctive crystals that formed on deposits of basaltic rock. Found mostly in crevices and hollows and generally modest in size, the crystals came in a variety of forms and colors that quickly made them prized objects for collectors. Their chemical composition was unusual and complex. They all seemed to contain significant amounts of aluminum and silica, but there were other chemicals present as well, and these seemed to vary.

In 1756, Swedish mineralogist Axel Cronstedt made a study of these crystals and discovered one with a surprising property: when exposed to a blowpipe's flame it rapidly swelled and expanded outward, producing a surface that under a lens resembled a mound of tiny, white milk-bubbles, which he described as "a white frothy slag." Further investigation revealed that several other crystals shared this quality, and from this and other distinguishing characteristics he declared them to be *sui generis*, a separate mineral species that he named "zeolites."

Much of the early scientific interest in zeolites came from their association with basalt, which was known to flow out of volcanoes as lava but also occurred in vast deposits where its origin was much less certain. Intriguingly, some basalt deposits were found to have small zeolite crystals scattered throughout them, while in other deposits these zeolites were completely absent. It was hoped that the presence or absence of these crystals might be an indicator of where (or even how) the basalt had originally formed—and that this, in turn, might shed light on the history of the Earth itself.

This hope was based on the work of the French mineralogist Faujas de Saint-Fond, who, in his book *Minéralogie des volcans* (1784), asserted that zeolites were the product of basalt's exposure to seawater and that "lavas that were not submerged, never contain it." In 1788, the geologist James Hutton employed this observation in his landmark work *Theory of the Earth*. He quoted Faujas, then observed: "Here would appear to be the distinction of subterraneous lava, in which zeolite and calcareous spar may be found, and that which has flowed from a volcano, in which neither of these are ever observed." A key part of Hutton's geological theory was that landmasses could rise and fall over time, and he used the fact that basalt containing zeolites could be found in the mountains around Edinburgh as evidence that at some point in the distant past they had been submerged. Smithson would have been intimately familiar with this line of reasoning, since his first scientific expedition had been to the Isle of Staffa with Faujas de Saint-Fond himself, and he met Hutton on that same trip.

Although they never actually yielded the hoped-for geologic insights, zeolites remained of considerable interest to mineralogists, and over the following years many new varieties were reported. But as the nineteenth century dawned there was still no consensus about what a zeolite was. German mineralogists, for example, used a mineral system based primarily on a mineral's physical appearance, and they used these properties to describe five separate types of zeolites: mealy, fibrous, radiated, foliated, and cubic. However, these categories proved difficult to apply and were not widely used on their own. There were also attempts to define zeolites based on their chemical composition, but the complexity of these minerals and the limitations of eighteenth-century chemistry made this a challenge. Some of the best chemists of the time worked on this problem, including Swedish chemist Torbern Bergman, Irish chemist Richard Kirwan, and Scottish chemist Thomas Thomson. But the only constant in their analysis of these minerals was the high proportion of silica and aluminum they contained, which was not considered sufficient to define them.

The more the zeolites were studied, the more confusing and arbitrary their identity became. By the early nineteenth century, when French crystallographer René Haüy came to study them, the situation had become so confused that he found it necessary to make a fresh start. The categories in his system were all based on each mineral species' crystals, and especially their angles. On that basis, he proposed that zeolites should become the name of a whole family of mineral species, and that the group of minerals which Cronstedt had identified as zeolites be renamed "mesotype." Under Haüy, the mesotype group became just one of the members of the new zeolite "family."

Haüy also announced that the mineral "natrolite"—which Cronstedt classified as a zeolite—had been misidentified, and he elevated it to the status of a unique species. Uncharacteristically, this decision was not based on measurements of the mineral's crystals. Natrolite's long, needle-like crystals were far too small to measure with the instruments that Haüy used. Visual comparisons of the crystals with those of other mesotypes suggested that they were similar, perhaps even identical. But Haüy needed more data to make that determination, so he turned to another source of information: natrolite's chemical composition.

Not being a chemist himself, Haüy consulted the scientific literature and found what he needed in the published analyses of two of Europe's most distinguished chemists: Martin Klaproth and Louis Vauquelin. In 1802, Klaproth had found that natrolite contained, in addition to silica and aluminum, a considerable amount of "soda" (sodium), while a few years earlier Vauquelin had found that minerals in Haüy's mesotype category contained "lime" (calcium), but no trace of soda. This distinction led Haüy to reclassify natrolite as a separate species, and it was this same distinction that, in 1811, prompted Smithson to write his article "On the Composition of Zeolite."

What Haüy and the European chemists did not realize was that in England it had long been an established fact that the zeolites—which Haüy now called mesotypes—contained sodium. As Smithson noted in his article, Hutton had found sodium in them back in the 1780s, and the Scottish chemist Joseph Black made this widely known by including it in his lectures. Smithson himself reported having found sodium in the zeolites he had collected on Staffa back in 1784, as well as in various zeolites that he had tested over the years. The presence of sodium in English zeolites was so widely accepted that in 1802, when the Scottish chemist Robert Kennedy finally put this fact in print, it hardly elicited any comment—which is probably why Haüy was unaware of it.

Indeed, Vauquelin's failure to find sodium in the samples he tested was so anomalous that Smithson doubted they were talking about the same mineral, and as it happened they weren't. The zeolites that Vauquelin tested had come from the distant Faroe Islands, an area now known to be geologically—and, in this case, mineralogically—distinct from the British Isles. Smithson also noted that it was not clear that any of the minerals tested by either Klaproth or Vauquelin "were of that species which Mr. Haüy calls mesotype."

Smithson likely communicated these doubts directly to Haüy, because around 1810 the French crystallographer sent him a package of mineral samples that included a cluster of regular crystals from the dormant Puy de Dôme volcano in central France. Labeled in Haüy's own handwriting, these were unmistakably specimens of Haüy's mesotype. As Smithson later wrote, they provided him with a "very favourable opportunity, to ascertain whether the mesotype of Mr. Haüy and natrolite, did or did not differ in their composition."

Smithson's chemical analysis of the crystals was thorough but not exhaustive. It was conducted, he admitted, "more for the purpose of ascertaining the nature of the component parts of this zeolite than their proportions," and in this he was eminently successful. He was able to show that Haüy's mesotype contained silica, aluminum, sodium, and water, just like natrolite, and in almost exactly the same proportions that Klaproth had found in natrolite. Smithson concluded that "the results of the experiments have been entirely unfavourable to their separation" and that mesotype and natrolite were simply variants of the same mineral species.

Smithson also argued in favor of retaining the name "zeolite" over Haüy's "mesotype." He wrote, "I think that the name imposed on a substance by the discoverer of it, ought to be held in some degree sacred, and not altered without the most urgent necessity for doing it. It is but a feeble and just retribution of respect for the service which he has rendered to science." This opinion would have important consequences for Smithson a few years later.

The British response to Smithson's article was almost universally favorable. "The process of the analysis was simple and satisfactory," wrote the reviewer for *Retrospect*, who went on to conclude that "the result of Mr. Smithson's analysis will go near to establish a perfect identity between these two minerals." Thomas Thomson also cited Smithson's analysis in his influential textbook, *A System of Chemistry*, noting that it showed natrolite to be "only a variety" of zeolite. He took Smithson's suggestion that the name "zeolite" be restored in favor of Haüy's "mesotype," which he stopped using. But, confusingly, Thomson still followed Haüy in grouping the zeolites, along with several other mineral species, as members of the "zeolite family"—a practice also followed by others.

Smithson's finding was also widely accepted in France, where it was revealed that Haüy had already been "convinced internally of the identity of the two substances" by his informal comparisons of each mineral's crystals but had felt compelled to distinguish between them because of Vauquelin's now-discredited analysis. Smithson's analysis was also cited in France as an example of the growing "conformity between crystallography and chemical analysis, which should raise hopes that the differences they sometimes present are only apparent, and disappear as the latter method more and more approaches perfection."

The attention Smithson drew to this topic also inspired other investigators, and in 1816 an exhaustive investigation of all the different varieties of Haüy's mesotype appeared in the *Journal für Chemie und Physik*. After three years of study, the authors concluded that the category (which, thanks to Smithson, now included natrolite) actually consisted of three separate species: one that contained soda, one that contained lime, and one that contained both soda and lime. And as part of their work, they also verified the analysis of "Herrn Smithson."

Smithson's article appeared in 1811, and within three years it had been translated into both French and German and reprinted in five other journals. This level of interest was rare for what was primarily a mineralogy article, but there was another aspect to the interest in zeolites, and that was based on the fact that they were composed largely of a class of matter called "earths."

"Earth" was the term used by early chemists to describe the inert, whitish, solid matter that sometimes remained at the end of a chemical analysis. Its identifying characteristics were puzzling and mostly negative: it did not dissolve well (if at all) in water, it was not strongly affected (if at all) by heat, and it was not strongly affected (if at all) by acids. So little was known about this material that it had long been an open question as to whether there were different kinds of earths or simply a single earth substance that acquired different properties when combined with other substances. It was only around the middle of the eighteenth century that chemists began to identify earths as a group of discrete materials.

By 1783 the Swedish chemist Bergman was able to identify at least five separate earths: terra ponderosa, calcareous earth, magnesia, argillaceous earth, and siliceous earth. Bergman described these naturally occurring substances as "natural compositions of acids with the earths," and although eighteenth-century chemistry was unable to isolate them, these compounds contained, respectively, the elements now known as barium, calcium, magnesium, aluminum, and silicon.

The study of earths became a major topic in late eighteenth- and early nineteenth-century science, and it was a subject that interested Smithson as well, evidenced by his determined search for a sample of terra ponderosa after visiting Staffa. He also investigated argillaceous earth, and around the same time that he wrote the zeolite article he made an unpublished analysis of a mineral that he identified as a "subsulphate of alumina." But his greatest interest was in understanding "siliceous earth," as can be seen in his work on tabasheer, in his identification of silica in one of the ores of zinc, and in his work on zeolites.

What made silica so interesting was that, although it was extremely common, it did not seem to follow the established rules of chemical combination. In Smithson's time, chemical combination was understood largely as the process of an acid combining with an alkali to produce a stable, neutral compound that chemists called a "salt." Acids did not combine chemically with each other, nor did alkalis, so if a substance was found to contain an alkali it must also contain an acid—and vice versa. Bergman's description of the compounds containing earths as being "natural compositions of acids" meant that the other component must be alkaline—which the earths all seemed to be, except for silica.

Silica did not react with most acids, but it did combine with hydrofluoric acid, which seemed to identify it as a base. It also combined with zinc (as Smithson had discovered in his work on calamines) and with other metals, which seemed to indicate that it was an acid. In nature it was frequently found in combination with other earths, as in the case of the zeolites, where it combined with "argillaceous earth," which also seemed to suggest that it was an acid. But silica had none of the other known properties of an acid, such as sour taste or the ability to change the color of indicator solutions.

Silica's "great analogy to acids" had been noted as early as 1788 in Axel Cronstedt's *An Essay towards a System of Mineralogy*—a book that Smithson owned. But it failed to meet the definition of an acid, and at that time the understanding of acid-base chemistry was so fixed that this line of reasoning was not pursued. Instead, investigators turned their attention to finding undiscovered acids in minerals containing silica, and it appears that Smithson himself was engaged in this search. In his article, he mentions corresponding with the Swiss chemist Henri Struve, who "was led to suspect the existence of phosphoric acid in several stones, and particularly in the zeolite of Auvergne, I have directed my enquiries to this point, but have not found the phosphoric, or any other acknowledged mineral acid, in this zeolite." The mechanism by which silica combined with other substances was so puzzling that there was even some suggestion "that bodies [containing silica] obeyed different laws, when brought together in the great laboratory of nature, from those which influenced them in our artificial processes." However, by the early nineteenth century, few chemists would have been willing to concede this point.

This was the confused state of knowledge in 1811 when Smithson wrote his paper on zeolites. It was his first publication since the disastrous sulphuret article nearly four years earlier, and he carefully limited his comments to the analysis of zeolites until the very end. There, with an economy of words that was unusual even for him, he made a sweeping observation: "Many persons, from experiencing much difficulty in comprehending the combination together of the earths, have been led to suppose the existence of undiscovered acid in stony crystals. If quartz [silica] be itself considered as an acid, to which order of bodies its qualities much more nearly assimilate it, than to the earths, their composition becomes readily intelligible. They will then be neutral salts, silicates, either simple or compound."

The suggestion that silica could act as an acid hardly seems remarkable today, but in 1811 it was a major conceptual leap. If what Smithson suggested was true, it not only challenged chemists' understanding of what an acid was but also identified a large new class of matter: the silicates. Accepting that silica acted like an acid also seemed to require changes to chemists' understanding of acid-base combination, which made the idea appealing to the Swedish chemist Jacob Berzelius.

Berzelius was in the process of developing an electrochemical theory of chemistry that explained chemical combination in terms of electrical attraction. He would later reveal that he had independently came to the same conclusion about silica as Smithson but hesitated to publish it "because the confused chaos which was spread over them [the silicates] might perhaps have had a tendency to prejudice the reader" against his new system. "I have since that period," he continued, "with a truly sincere pleasure, learned that Mr. Smithson, one of the most experienced mineralogists of Europe, without any knowledge of my essay, has published a similar idea in a memoir on the nature of stilbite [zeolite] and mesotype." Emboldened by Smithson's confirmation of his suspicions, he then went on to spend fourteen pages discussing the various types of "silicates" that could be explained by this assumption.

In the system that Berzelius proposed, acids were "electro-negative" and bases were "electro-positive." In reactions where silica combined with a base, the fact that opposite charges attract explained their combination. But in cases where two acids combined, such as silica and hydrofluoric acid, a different rule applied. In these situations, the electrical strength of the acids was of crucial importance, and Berzelius informs us that "in the union of two acids the weaker acid serves as the basis to the stronger." Depending on the situation, silica could act like either an acid or a base.

Berzelius was not the only one who adopted Smithson's suggestion. Thomas Thomson noted it in his review of Berzelius's book and later incorporated it into his own textbook, writing, "Silica is the most common ingredient in stony bodies, and exists in them, combined with various earths and metallic oxides. Mr. Smithson suggested that in these compounds the silica performs the function of an acid; on opinion which has been demonstrated in a satisfactory manner by Berzelius." Thomson considered this to be a fundamental insight, and he noted nearly twenty years later in his *History of Chemistry* (1831), "This great discovery, which has thrown a new light upon mineral bodies, and shown them all to be chemical combinations, was reserved for Mr. Smithson."

Even the English chemist Humphry Davy, whose many discoveries often overshadowed the work of others, gave Smithson credit in several of his articles. In one he wrote that "the facts brought forward by Mr. Smithson and M. Berzelius, tend to show the propriety of arranging silica in the class of acids," and in another that "I think Mr Smithson is the first person who made this statement & long before Berzelius." Smithson was also credited in France, where translations of several of Davy's articles appeared in French journals, and where the *Dictionnaire de chimie* remarked on the "ingénieuse observation de M. Smithson."

Smithson's suggestion was not received entirely without opposition. William Brande, Humphry Davy's successor as professor of chemistry at the Royal Institution, took great exception to Berzelius's theory, calling it "a chaos of inverted reasoning and gratuitous assumption," and he mocked Thomson for supporting it. His criticism of Smithson was somewhat less vitriolic but still highly critical, arguing that if silica, which in nature existed as an oxide, could act as an acid, then other oxides should have the same property.

Most of the other objections related to the change in the definition of an acid. The chemistry textbook author John Murray wrote that there was "nothing in the chemical agency or combinations of silica to justify this opinion; no body is more remote in its properties from acidity, and it is to confound all chemical distinctions, to associate it with substances totally different." Another chemistry-text writer, Franklin Bache, reported "not being satisfied that its close analogy to acids has been well made out. There are indeed some circumstances with regard to silica, in which it differs particularly from all the other earthy salifiable bases; but these do not call for the measure of associating it with acids." Substantial as these arguments were, they did not stand for long. In 1822 Thomas Thomson noted that Brande had stopped criticizing the idea of silica acting like an acid, and that it could no longer be doubted by "any well informed chemist."

Berzelius's theory would be superseded in the following years but, despite that, the idea that silica acted as an acid remained a permanent fixture in chemical and mineralogical thinking. Smithson's observation came at a particular historical moment. Had he not made it, someone else soon would have; indeed, at least two other individuals have been identified as having made the same observation around the same time. But neither of them had Smithson's scientific stature, which was what was required to earn consideration of such a radical new idea. Conversely, the fact that simply throwing out such a simple remark could have received so much attention speaks to Smithson's reputation and to how closely his articles were read.

During his lifetime, Smithson was widely credited for having made an important contribution to both chemistry and mineralogy, but by the middle of the nineteenth century the memory of that contribution began to fade. In 1862, the secretary of the French Academy of Sciences publicly, and mistakenly, credited the new understanding of silica to the English chemist Smithson Tennant. Several years later that misattribution was repeated in the *Annual Report* of the Smithsonian Institution—the very institution that Smithson had left his fortune to establish.

———————+———————

8.

ULMIN

JAMES SMITHSON'S SIXTH ARTICLE, "On a Substance from the Elm Tree called Ulmin," was published in 1813. His previous writings had been concerned with the chemistry of minerals. Even his article on tabasheer had been about a mineral—albeit a mineral found in bamboo plants. But now he turned his attention to a fundamentally different type of material, one related to the processes of life.

Today the substances produced in living tissue are studied with the concepts and techniques of what we call organic chemistry. In Smithson's time this branch of chemistry did not yet exist, and what could be learned from these kinds of materials was limited. The techniques of high temperature and acid-base analysis that chemists used to study minerals were destructive to organic samples, and beyond confirming that they were composed almost entirely of carbon, hydrogen, nitrogen, and oxygen, these techniques revealed little. Instead, chemists had to limit themselves to studying these materials with less destructive techniques like extraction, the process of isolating materials by dissolving them. In Smithson's time this was most commonly done with water, but extraction could also be done with other liquids, such as alcohol. Chemists of the time had to content themselves with only being able to identify chemical "principles"—substances within the tissues of plants or animals that seemed to be fundamental to their constitution. Ulmin, the material Smithson now chose to study, had recently been identified as one of these "principles," and as a consequence it had become a topic of considerable interest.

The story begins with the French chemist Louis-Nicolas Vauquelin, whose analysis of zeolite was discussed in the last chapter. Vauquelin was a pioneer in the development of methods for analyzing plant materials. Several years earlier, on a visit to the Parc de Saint-Cloud in Paris, he noticed a liquid issuing from wounds on the trunks of elm trees and was struck by the similarity of these injuries to the "running ulcers" sometimes found on animals. He collected samples of bark that had been soaked by the liquid, and in 1797 he published his analysis.

Vauquelin reported finding two types of liquid running down the trees' bark. The first was a clear liquid that collected on the bark around and below the wound, creating an appearance "like a white limestone." The second was a dark-colored secretion that soaked into the bark immediately below the wound, turning it a shining black color, "as if one had applied a varnish." He reported that both were water-soluble, slightly alkaline, and contained potassium, but it was the matter that he extracted from the dark-colored bark that seemed to interest him the most, and that is the substance referred to in this discussion. He described soaking the bark he collected in water, which extracted a dark-colored liquid that, when filtered and left exposed to the air, slowly thickened to a blackish-red syrup. Unlike many materials extracted from plants, this syrup did not seem inclined to mold, and he reported that it had an alkaline taste, a finding he confirmed by noting that, when put in contact with an "indicator" solution made from turnsole, the annual herb *Chrozophora tinctoria*, it changed the color from red to green.

Vauquelin's experiments with this syrup fell roughly into two groups. In one group he tested the syrup to see how it fit into the categories of known plant principles, but the results of these tests were ambiguous. It readily dissolved in water, but not in alcohol, properties which were qualities of a "gum." But he also found that adding dilute sulfuric acid to the syrup caused it to produce reddish particles that slowly settled to the bottom of the container. Vauquelin collected these particles and reported that water would no longer dissolve them, which seemed to put them in the category of a "resin." He also tested the liquid from which the particles had been removed and found that it contained potassium, which explained the syrup's alkalinity.

The second group of tests explored the similarity of the syrup to the fluid produced by an animal ulcer, and Vauquelin reported that although alcohol failed to dissolve the syrup, it did produce a surprising effect. Adding alcohol caused the syrup to form a "very thick ropy coagulum," which seemed to resemble mucus. When he added metallic oxides to the syrup (the application of metallic oxides being a common treatment for various human ulcers), they changed it into a thick paste.

Vauquelin's findings were inconclusive but provocative, and for him at least, they suggested the possibility of creating "medicines" for plants, similar to those used on animals. However, the unresolved nature of the elm syrup—whether it was a gum or a resin or something new and completely different—lingered. The question was soon addressed, though. In 1800 Smithson's friend William Thomson was in Palermo, Sicily, where like any good scientific traveler he spent part of his time collecting interesting plant and mineral samples to send to his friends. One of the things that caught his eye was the extrusion of a "saline gum from an old elm tree," and he sent samples of it to Smithson and the German chemist Martin Klaproth.

Klaproth's report on it was delivered two years later. Although he was one of the most respected chemists in Europe and a pioneer in quantitative methods, his "analysis" was entirely descriptive and failed to list a single measurement. In fact, there were not yet many meaningful measurements that could be made in plant chemistry, and Klaproth seemed to acknowledge this in his opening statement: "Although the progress of science has increased the number & complexity of the theorems regarding the chemistry of vegetation, this branch of human knowledge has not yet developed enough to allow one to neglect any detailed observation reported about it."

Unlike Vauquelin, whose analysis he seems to have read, Klaproth chose not to comment on the possibility of an analogy between plant and animal diseases. Instead, he limited his remarks to the question of how to classify this material. He also chose a different starting point for his investigation. Where Vauquelin had worked with a syrup—a thickened version of the material he extracted from the elm bark—Klaproth let the extract dry completely into a shiny, dark brown solid that he then ground into a powder. This was the starting point for all his tests, as it would be for every investigator who came after him.

Klaproth tried dissolving the powder in both alcohol and ether, but nothing happened. The powder was unaffected and did not even change the color of either liquid. This ruled out the possibility of its being a resin, since resins, by definition, were dissolved by alcohol. By contrast, the powder was easily dissolved by even a small amount of water, which was one of the main characteristics of a gum. There were other tests that also supported this conclusion, but probably the most critical quality of a gum was, as the name implies, that it be adhesive, and Klaproth reported that this material "differs absolutely from the gums, by the total absence of being sticky."

Klaproth observed that a solution of the powder had an extremely bland taste and that a large residue of charcoal was left after the powder burned. He confirmed Vauquelin's finding that the solution was alkaline and that it contained "potash" (potassium). He also tested it with a small amount of acid and found, as Vauquelin had, that it caused particles to settle on the bottom of the container. Allowing the solution to evaporate left a reddish-brown powder, similar to what Vauquelin had described. Klaproth then performed one more test and made a surprising discovery. Where alcohol had previously had no effect on the powder, it was now able to partially dissolve it. And where water had originally dissolved it easily, it now seemed to have no effect. Adding a little acid to this material had exposed an entirely new property, changing it from first being soluble in water (but not in alcohol) to being partially soluble in alcohol (but not in water). It was almost as though a gum was being converted into a resin. This was so unexpected—the only class of materials showing any resemblance to this property were the "volatile" (or aromatic "essential") oils—that Klaproth concluded: "The solidified juice of the elm discussed here, does not appropriately belong to any of the categories to which we assign the immediate products of plants."

It fell to Scottish chemist Thomas Thomson to take the next step, which he did in his influential textbook *A System of Chemistry* (1807). He described the "very singular substance lately examined by Klaproth," and declared that it "differs essentially from every other known body, and must therefore constitute a new and peculiar vegetable principle." And he finally gave this substance a name. He called it "ulmin," from the Latin *ulmis* (elm). Thomson put ulmin in the "extractive" class of plant substances, which were "soluble in water, at least in some state or other, and which in general are solid, and not remarkably combustible." Resins, by contrast, were in the class of "oleoform" materials, substances that were "either fluid or which melt when heated, and burn like oils. They are all insoluble in water, but in general they dissolve in alcohol."

Thomson initially relied on Klaproth's analysis to describe the properties of ulmin. However, as he noted, the study of plant materials was "a branch of chemistry by no means remarkable for its precision," and when one of his readers sent him a sample of what appeared to be an ulmin "exudation from an old elm in the neighbourhood of Plymouth," he took the opportunity to make his own investigation, which appeared in early 1813. The material he extracted from the sample exhibited many of the qualities described by Klaproth, but when he added acid to a solution of it, the resulting precipitate failed to show the expected effect. He reported that it was "not sensibly soluble either in water or alcohol," and concluded

Ingredients of Plants

I. Extractive	II. Oleoform	III. Fibrous	IV. Extraneous
1. Acids	1. Fixed oil	1. Cotton	1. Alkalies
2. Sugar	2. Wax	2. Suber	2. Earths
3. Sarcocoll	3. Volatile oil	3. Wood	3. Metals.
4. Asparagin	4. Camphor		
5. Gum	5. Bird-lime		
6. Mucus	6. Resins		
7. Jelly	7. Guaiacum		
8. Ulmin	8. Balsams		
9. Inulin	9. Gum resins		
10. Starch	10. Caoutchouc		
11. Indigo			
12. Gluten			
13. Albumen			
14. Fibrin			
15. Bitter principle			
16. Extractive			
17. Tannin			
18. Narcotic principle			

In Thomson's list of the known plant substances, ulmin is listed as the eighth "extractive" substance. This list appeared in the fourth edition of his widely read book *A System of Chemistry* (1810), the same edition that Smithson read and that inspired him to make his own analysis of ulmin.

that "the characters ascribed to ulmin by Klaproth do not apply to the substance which I examined."

It was a puzzling development, and it seemed there were two possible explanations. Either Thomson or Klaproth had made a mistake, or there was some difference in their samples. Because one sample had come from an English elm and the other from a Sicilian elm, it seemed likely that the trees were different species and perhaps had a different chemistry. Simply retesting both samples in a side-by-side comparison should have determined if this was the case, but Thomson did not have a sample from a Sicilian elm, and his prospects of getting one seemed remote.

Fortunately, there was one man in England who did have such a sample—James Smithson. He still had the sample that his friend William Thomson had sent him, which was from the very same tree that had produced Klaproth's sample. And Smithson, who apparently had been following the discussion in the literature, had already begun an analysis of it. He and Thomas Thomson were moving on parallel courses. Thomson's article on ulmin went to press in December 1812, and just a few days later Smithson's own analysis of the material was read to the Royal Society. Printed in early 1813, it was more detailed and complete than anything yet published. He explained how he had extracted the ulmin from his sample by soaking it in water, filtering the resulting solution and then drying it to a solid in a "water bath." He described how lumps of the ulmin appeared black, but that thin pieces were "transparent, and of a deep red colour." A dilute solution was yellow, but a concentrated one was "dark red, and not unlike blood." The solution was slightly alkaline, and when dried in a watch glass it formed a distinctive pattern that he described as "long strips disposed in rays to the center."

Where previous investigators had tested their ulmin solutions with either one or two different acids, Smithson tested his with seven, each of which produced a precipitation of "the same resin-like substance." He also reported that the liquid left after filtering each of these solutions invariably contained potassium, which he measured as being roughly one-fifth of the total weight of the ulmin that produced it. With regard to the properties of the resin-like substance, Smithson reported that both alcohol and water dissolved it, "but only in very small quantity." He also found that the solution in water was slightly acidic and, significantly, that adding a small amount of potassium to water with a bit of solid resin in it caused the resin to dissolve "immediately and abundantly." He reported that the resulting liquid "has all the qualities of a solution of ulmin," including that when it dried, it formed the characteristic pattern of rays that he observed earlier.

This was a new discovery, and it must have been a particularly satisfying one for chemists. Smithson had been able to break ulmin down into two components—the "resin," which was slightly acidic, and potassium, which is an alkali—and then he had recombined them to produce ulmin. This kind of manipulation was common in other branches of chemistry but relatively rare in plant chemistry. Smithson's analysis clarified the nature of ulmin's properties and identified a potentially important new plant "principle."

Smithson also tested a substance he had collected from a diseased English elm growing in London's Kensington Gardens. He found that it had all the properties of ulmin and that, as Thomson had reported with his sample of English ulmin,

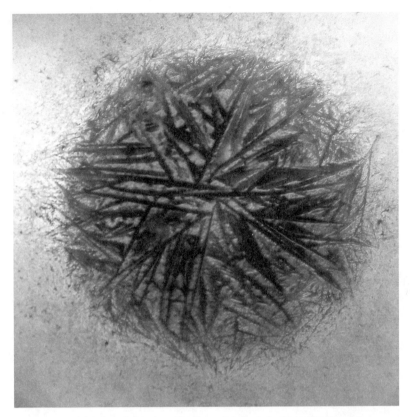

The pattern of dried ulmin produced by the author's replication of Smithson's experiment. It closely resembles Smithson's description of "long strips disposed in rays to the center," which he provided as an alternative way of identifying ulmin. As Klaproth noted earlier, plant chemistry was not yet to the point where any detailed observation could be ignored. The bark used to prepare this solution came from a diseased elm tree near the Smithsonian Castle in Washington, DC. Photograph by Steven Turner.

the material precipitated by acids was not dissolved by alcohol—a property that he attributed to the English ulmin's having a higher potassium content.

Finally, with regard to the question of how ulmin should be classified, Smithson deferred on this point, declaring simply that his analysis showed that ulmin was not a vegetable principle of "anomalous qualities." He seemed to feel that the material of real interest was the acidic matter within ulmin, which he declared to be "rather more related to the extractives than to the resins." He did not name the material, but it would later come to be known as "ulmic acid."

Although Smithson's analysis effectively refuted the claims of Klaproth about ulmin's unique properties, he deliberately refrained from using the German chemist's name when he made this point. Indeed, in the manuscript that Smithson submitted for publication, a footnote mentioned Klaproth with warmth and admiration. Those comments, however, were deleted by the editor and were never published.

Plant chemistry was a topic of considerable interest in the early nineteenth century, and Smithson's paper was widely read. Summaries of it were quickly printed in the *Annals of Science* and *Philosophical Magazine*, and the latter journal soon reprinted the entire article. Detailed accounts of it also appeared in *The Monthly Review* and *Retrospect*, the latter of which commented: "This paper of Mr. Smithson's consists of the full particulars of a chemical analysis which will hardly admit of abridgement." It was also mentioned in foreign journals, and the entire article was translated into French and published in the *Journal de Physique*.

Thomas Thomson, whose 1810 book had inspired Smithson's analysis, also acknowledged the article, noting that because his sample had come from the same tree as Klaproth's, "this circumstance renders Mr. Smithson's observations peculiarly valuable." Thomson incorporated Smithson's findings into the discussion of ulmin in the next edition of his *A System of Chemistry* (1817), remarking that "these experiments were made by Mr. Smithson, well known for his precision." Smithson had made a timely and important contribution to a topic of wide interest, an achievement for which he was justifiably proud. As others investigated further, ulmin was soon seen as a key component of both plants and the soil.

Spurred on by his readers, who continued to send him samples of dark-colored matter collected from different types of trees, Thomson wrote several more articles on ulmin. He found that the amount of potassium in the ulmin of different types of trees varied widely and that a sample from an oak tree "turned out, on examination, to be ulmin quite free from potash [potassium], a substance which had constituted a part of all the specimens formerly examined either by Mr. Smithson or myself."

At around this same time, the Swedish chemist Jacob Berzelius reported having found that ulmin was a constituent of the bark of almost every type of tree. According to him, this had gone unnoticed because the usual method of extracting ulmin, which was to soak the bark for a long time in hot water, often caused the ulmin to combine with tannin and become undetectable. Berzelius got around this by first soaking the bark in hot alcohol and then in cold water, after which the ulmin could be extracted in the usual manner with hot water.

A few years later, French chemist Henri Braconnot reported finding ulmin in a wide variety of vegetable-based materials, and he described a method of making

it artificially from sawdust. His long article, "Memoir on the Conversion of ligne-
ous Fibre into Gum and Sugar, and into an Acid of a peculiar nature, by means
of Sulphuric Acid; and on the Conversion of the same into Ulmin by Means of
Potash," was subsequently translated and reprinted in *The Repertory.*

Ulmin was reported in the most unlikely places, such as in the roots of the
asarum plant (wild ginger) and in coal. Even cast iron was found to contain a trace
amount of it—apparently an artifact of the coal used in its manufacture. Ulmin
seemed to be everywhere, particularly in the soil. In 1831 the *Journal of the Royal
Institution* called it "the most valuable compost known" and opined that "Holland
probably owes the superiority of its agricultural productions to the quantity [of
ulmin] which it naturally possesses." Ulmin (or "ulmic acid") was seen by many
as the vehicle by which plants drew nutrition from the soil, and Smithson's role in
identifying it was frequently acknowledged.

But this perception of ulmin's importance was undermined by a growing suspicion that it was greatly overstated. Despite important advances in the chemical
analysis of plant materials, the definition of ulmin being used in the 1830s was precisely the same one that Smithson had proposed nearly two decades earlier—and
it was too general. William Brande declared that "ulmin is of a dark brown colour,
with scarcely any taste or smell. It is sparingly soluble in water and in alcohol, but
readily soluble in a weak solution of carbonate of potassa [potassium]." In practical terms, this meant that any insipid brown liquid that could be extracted from
plants or the soil had a good chance of being called ulmin. Finally, in 1832, Berzelius called for a halt to the term's use, arguing that it had been applied to "a host
of substances that, despite their superficial similarity, can exhibit very different
chemical properties." "It is inaccurate," he wrote, "to join together under a common name, little-known materials which present some analogies. This abuse sufficiently allows rejection of the word ulmin."

Instead of ulmin, Berzelius proposed the name "geigne," from the Greek *ge*
(Earth). His purpose in introducing this new term was to stimulate a new investigation of the components of soil, one based on chemical composition rather than a
search for chemical principles. But although chemists could now accurately measure the proportions of the elements in these substances, they still had no way to
think about their structure, which is critical to the modern understanding of how
they work. That understanding was still far in the future, and rather than bring
clarity to the discussion, Berzelius's rejection of "ulmin" turned out to be the first
shot in what became a long, confusing battle among chemists about the nature of
soil. As one observer later remarked,

Some writers speak of the generic substance under the name of humus, some under the name of mould, and some under the name of geine; some regard as its principal species or modifications ulmin, humic acid, coal of humus, and humin,—and others, ulmin, ulmic acid, humin, and humic acid; some include among its species or modifications crenic acid and apocrenic acid, and others deny the very existence of these acids; and while some differ from others in theory, in definition, or in analysis, others differ chiefly or solely in mere nomenclature.

By midcentury, the argument had led many practical men to throw up their hands in frustration. "If farmers are to wait till doctors agree, there will be no harvest," wrote the American agriculturalist Samuel Dana. "It is a difference about names, not things."

It would be many decades before organic chemistry finally developed to the point where it could speak with authority on soil and plant chemistry. By that time Smithson's work, along with the rest of the history of ulmin and most of the history of "vegetable chemistry," had long since become irrelevant. Smithson had made what was considered to be an important contribution at the time, and his work had never been shown to be incorrect, but as part of a branch of chemistry that came to be seen as a dead-end, it was effectively pruned from the scientific record.

———————+———————

9.

A SALINE SUBSTANCE
FROM MOUNT VESUVIUS

IN 1813, James Smithson finished his seventh article. Titled "On a Saline Substance from Mount Vesuvius," it was only six pages long, but it is unique among his
writings. Instead of the brief factual introduction that he usually provided, this
article starts with a series of surprisingly broad assertions about geology. The first
is that at some point in the past Earth had been "in a state of general conflagra-
tion," and that it is probably "an extinct comet or sun." He acknowledged that this
suggestion was not new and had previously been difficult to prove, but that "great
modern discoveries" had dissipated those difficulties. He did not name the discoverer (or provide any other details), but credited the knowledge "that the bases of
alkalies and earths are metals" with providing an explanation for volcanoes and
their products.

The argument is rushed and not fully developed, but Smithson goes on to
assert that the reason the Earth is no longer burning is that, at some point in the
past, a crust—a "calcining mass"—formed over it and extinguished the fire. That
crust is now the "stony envelope" on which we live, but Smithson wrote (without
any supporting details) that a "metallic ball" of "combustible matter" still exists
underneath it, and that this is what causes volcanoes. He concludes by noting the
importance that this theory places on volcanoes and their products, since if it is
true "they cease to be local phenomena; they become principal elements in the

history of our globe; they connect its present with its former condition; and we have good grounds for supposing, that in their flames are to be read its future destinies."

It took Smithson only five paragraphs to say this, and to support it he offered just one observation: that the "primitive strata," the layer of rock then thought to be Earth's oldest, had never yielded crystals with water inside, and that "none of the materials of these strata contain water in any state." Smithson then abruptly assumes a different tone and begins to describe his analysis of a curious rock. It had flowed from an opening in the Mount Vesuvius volcano, and was collected and sent to him by his friend, William Thomson, nearly twenty years earlier. After a typically precise chemical analysis, Smithson concludes rather cryptically that "this Vesuvius salt, considered in its totality, has presented no less than nine distinct species of matters, and a more rigorous investigation, than I was willing to bestow on it, would probably add to their number."

Historians have shown little interest in this article for a number of reasons. In the introduction Smithson fails to provide a context for this study, the remarks on geology seem overly broad, and he fails to provide any references. By contrast, while the analysis that follows is precise and does contain references, it comes to no specific conclusion. Even worse, Smithson appears to indicate that it isn't quite complete.

However, this article is not so easy to dismiss. Smithson was a widely respected natural philosopher; his article was read to and approved by the Royal Society and it was published in England's premier scientific journal, *Philosophical Transactions*. Moreover, volcanoes and their products were items of particular interest in Smithson's time, all of which leads to the suspicion that part of the story has been lost or overlooked.

In the article, Smithson's praise of the "great modern discoveries" about the metallic bases of alkalis and earths stands out. In view of his other writings, it is unlike him not to give a reference for this kind of remark, but a little research shows that this is a clear reference to the work of Humphry Davy, the English chemist who, just a few years earlier, had used powerful batteries and a dramatic process he called "electro-chemical analysis," to discover seven new chemical elements.

Chemical batteries had been in existence for only a few years when Davy started his experiments. Alessandro Volta is credited with building the first one in 1799, and in the following years a great deal of time and effort had been expended to increase their power. The chemical batteries produced a steady flow, or "current" of electricity, which is very different from the "static" electricity that

112

naturalists had traditionally studied. Static electricity manifested itself mostly as electric sparks that, while powerful, lasted only a moment. The electricity of batteries, on the other hand, while not inclined to make great sparks, had a more important property—an increased ability to decompose matter. Davy's experiments used this property to go beyond ordinary chemical analysis and to achieve, as he put it, "a more intimate knowledge" of some important chemicals.

He initially used the three large voltaic batteries of the Royal Institution, which, by the standards of the day, were impressive. The largest consisted of more than two hundred sets of six-inch square copper and zinc plates, activated by a saline solution. When used in series with the other two batteries, they could produce (in modern terms) approximately 200 volts of direct current, all of which Davy directed into his small samples—to great effect.

In 1807 he described using his apparatus to analyze a small lump of carefully dried "potash" (potassium hydroxide). Placed between the battery's terminals, the parts of the potash exposed to the current soon began to melt. At the positive terminal, oxygen bubbled up to the surface, while at the negative terminal Davy described how "small globules having a high metallic lustre, and being precisely similar in visible characters to quicksilver [mercury], appeared, some of which burnt with explosion and bright flame, as soon as they were formed, and others remained."

Davy collected these remaining globules and later confirmed that they were "basis of potash"—the never-before-seen material potassium. He had to work quickly; potassium is extremely reactive, and the small shiny globules would begin to tarnish as soon as they formed. If left exposed to the air for any length of time, they would rapidly combine with oxygen and soon would be entirely covered with a white film. The new metal was so reactive that no matter how carefully it was stored, it would eventually combine with something and be lost. Davy did find that putting potassium in a jar of highly distilled liquid naphtha would preserve it for several days, but no longer.

Later the same year, Davy performed an "electro-chemical" analysis on another material, caustic soda (sodium hydroxide), and once again discovered a new element that had never been seen before. He called this one "basis of soda," now known as sodium. It was a solid at room temperature, while his potassium was liquid. Like potassium, this new metal was extremely reactive.

Both substances were particularly reactive to water, a property that Davy liked to demonstrate by tossing a small bit of each metal into a bowl of water. Potassium is so reactive that it doesn't sink into the water but remains on the surface, scurrying back and forth, supported and propelled by the release of a small cloud of hydrogen gas and accompanied by a distinctive violet flame. The reaction of sodium was even more dramatic. The instant it touched the water it would explode, leaving behind only a few waves on the water and a small ring of cloud that would expand and dissipate as it drifted slowly upward.

Davy's discoveries were so important and so numerous that for an unprecedented five years in a row he was asked to give the Royal Society's prestigious Bakerian lecture. These talks were subsequently published in the *Philosophical Transactions*, and Davy was widely seen as having revolutionized the science of chemistry. It seems clear that these were the "great modern discoveries" that Smithson referred to in his article. But what did all this have to do with volcanoes and geology?

The answer is that, like many of his scientific colleagues, Davy had a deep and long-held interest in geology, and, like Smithson, was inclined to support the idea that heat had played an important role in forming the Earth. Chemistry and geology were widely seen as related topics, and the Royal Institution committee on which they both served was named the "Chemistry and Geology Committee." An integral part of Davy's chemical research was exploring what his discoveries might mean for the scientific understanding of the Earth.

In particular, his discovery of the alkaline metals potassium and sodium suggested a theory that made a lot of sense. In the late eighteenth century, just a few years earlier, Henry Cavendish had used extremely accurate measurements of Earth's density to make a strong case for its having a metallic core. Building on this, Davy reasoned that if the reactive metals he discovered were present in that core, then their existence offered a powerful new explanation for volcanoes, lava, and the subterranean heat. If potassium and sodium existed in a pure metallic state just beneath Earth's crust, then volcanoes might be produced when accidental openings in that crust allowed water and air to come into contact with them. Instead of being the result of massive underground fires, always a problematic idea, Davy was suggesting that volcanoes were the result of a violent chemical reaction.

Davy's theory explains Smithson's puzzling reference to a "metallic ball" under the earth's crust, as well as his subsequent reference to a "large body of combustible matter" that was responsible for volcanoes. But if this is what Smithson was writing about, why didn't he refer to Davy by name or provide any references to his theory? This is where the story gets complicated, because in England during the early nineteenth century, proposing a new theory about volcanoes could be dangerous, despite the particular fascination that the English had long had with them.

To be sure, volcanoes have long been a topic of interest to all European peoples, both in literature and natural philosophy. However, beginning in 1764, and continuing for several decades, English interest in volcanoes increased significantly. That was the year William Hamilton, Britain's new envoy to the Spanish court, arrived in the Kingdom of Naples. While this was not a prestigious post at the time, Hamilton requested it in the hope that the climate would restore his wife's health. His assignment to Naples also furnished him with an income and a residence, and he soon discovered that his few official duties were easily fulfilled, leaving him with abundant free time. By chance, his arrival coincided with the beginning of a long period of activity for the city's resident volcano, Vesuvius. He began to keep a record of its activities almost as soon as he arrived, and his interest soon became a passion. For the next thirty-five years he would continuously study the volcano and send regular reports about it back to England. Many of his reports went to the Royal Society, which published them in its journal and made Hamilton a member. The reports were also collected into widely read books. But this does not convey the full scope of what he did, because Hamilton was in a position to apply not only his intellect and energy to the study of volcanoes, but his wealth and taste as well.

To begin with, Hamilton acquired a number of villas in the area around Vesuvius, which provided him with comfortable and relatively safe places to view whichever side happened to be active. He also organized and funded collecting expeditions, documenting the volcano and the substances it produced on a scale that had never before been attempted. He even commissioned artists to create a record of its changing appearance, and many of their drawings were later made into prints and paintings that Hamilton displayed on his trips back to London. Some of those images also found their way into Hamilton's magisterial *Campi Phlegraei* (1776), a book that combined his engaging reports with more than fifty hand-colored prints.

Not surprisingly, Hamilton's reports made many readers want to experience a volcano themselves, and he accommodated this by commissioning several replications in the form of paintings made with what were called "transparent colours." One was donated to the Royal Society, and another was displayed at the British Museum. There, situated in the middle of a darkened room with lamps placed behind it on an elaborate clockwork mechanism, the painting was said to give the perfect illusion of a current of glowing lava.

Hamilton was not alone in making replica volcanoes. The continuing interest in these natural wonders created an opportunity for London's "pleasure gardens." Sometimes described as Georgian amusement parks, at least one of them usually had a volcano replica on display until well into the nineteenth century. The Marylebone Gardens, for instance, offered the "Forge of Vulcan," a reproduction of the side of Mount Etna in which the mountain appeared to erupt and lava rushed down its side. In the 1780s, Covent Garden had a similar attraction, which "erupted" every two hours, six times a day. The "lava" in these displays was generally water that ran in troughs covered with red glass, illuminated from behind by lamps. Indoor replicas were also made, many using burning corks to simulate lava, with loud sound effects and burning sulfur added to heighten the experience.

This was the context in which Davy developed his ideas about volcanos and in which, from 1805 until 1811, he presented two annual lecture series at the Royal Institution; one about chemistry and the other about geology. Both courses were well attended, with as many as a thousand people packed into the institution's theater, and Davy used them to describe his ongoing research and speculate on its broader significance. Like most of Davy's talks, these lectures had demonstrations, which included model volcanoes. These were particularly engaging because while the commercial replicas that Londoners were used to seeing were intended to give just the experience of a volcano, Davy's models proposed to demonstrate the actual mechanism by which real volcanoes were formed.

He appears to have made several different models, their details changing from year to year to incorporate his new ideas and discoveries. His first model was based on the suggestion that volcanoes were the product of the spontaneous combustion of mineral pyrite, which under certain circumstances was known to burst into flame and leave a glowing residue somewhat similar to lava. Not many details are known, except that the pyrite was mixed with coal dust and an oxidizing salt and was ignited by a drop of acid. The most elaborate model Davy made was the one he used in his final geology course in 1811. It was a three-dimensional clay replica of Vesuvius, and inside some of its crevices it contained metallic potassium, iron, and calcium to simulate the Earth's metallic core. It also seems to have been connected to an electrostatic generator to simulate lightning. The model was activated by the simple act of adding water, and the effect was spectacular. As one eyewitness later reported, "the metals were soon thrown into violent action—successive explosions followed—red hot lava was seen flowing down its sides, from a crater in miniature—mimic lightnings played around: and in the instant of dramatic illusion, the tumultuous applause and continued cheering of the audience [filled the auditorium]."

Davy's geology lectures were a triumph, and his ideas about volcanoes are an important part of this story, but he only mentioned them briefly in his writings. Indeed, throughout this period Davy's most detailed account of his theory consisted of a single paragraph near the end of his 1808 article: "The metals of the earths cannot exist at the surface of the globe; but it is very possible that they may form a part of the interior; and such an assumption would offer a theory for the phaenomena of volcanoes, the formation of lavas, and the excitement and effects of subterraneous heat, and would probably lead to a general hypothesis in geology." This was supplemented by a short footnote that read: "Let it be assumed that the metals of the earths and alkalies, in alloy with common metals, exist in large quantities beneath the surface, then their accidental exposure to the action of air and water, must produce the effect of subterranean fire, and a product of earthy and stony matter analogous to lavas."

Until 1828, those two sentences were the most complete statement of his theory that Davy put into print, although he seems to have described it at length in his talks. In a review of his 1811 geology lectures, *Philosophical Magazine* reported that Davy devoted the last lecture entirely "to the consideration of the causes and effects of volcanos," and it described how he discussed in detail each of the points that Smithson would later make in the beginning of his article, including Davy's suggestion that "the consideration of volcanic fires relate to the future order of things."

Humphry Davy was actively advancing a geologic theory that he chose not to publish. He was so interested in it that twice, in 1814 and again in 1819, he visited Vesuvius for extended periods, conducting experiments on both occasions in the hope of finding evidence to confirm his theory. Yet he never went to the trouble of writing it out and putting it in print. This was not how science normally worked, and it was certainly not how Davy normally worked. His willingness to speculate was what had made him a scientific celebrity, and this theory seemed to be of interest to everyone. What could have caused his strange reluctance? The answer lies in the climate in which science was practiced in England, and how the French Revolution caused it to change.

Throughout most of the eighteenth century, most educated English men and women saw science through the lens of what one historian called "that particularly English phenomenon, the holy alliance between science and religion." This view was part of a larger worldview that historians have come to call "the Christian Enlightenment," a set of beliefs widely held among the English upper classes, which found its ultimate expression among the royal family and their inner circle at Windsor. Almost exclusively Protestant, "Enlightened" Christians have been characterized as having "a relatively optimistic view of human nature and a generally positive attitude towards both reform and progress," as practicing "a simpler, clearer, more tolerant and morally efficacious religion," and as believing "that their religion would be strengthened, not weakened, by its association with science and reason." King George III and Queen Charlotte were widely seen as models of these values.

One of the tenets of the Christian Enlightenment was that divine revelation could be confirmed by the study of the natural world. This had long been a fixture of English science, particularly in geology, but it had found its fullest expression in the work of Jean André de Luc. Born in Protestant Geneva, another corner of the Christian Enlightenment, de Luc was a widely respected natural philosopher who had come to England in 1784 to be the "scientific reader" to the Queen. Charlotte, like King George, had a deep interest in science and believed that science should support the teachings of the Bible. De Luc was also deeply religious, and this, in addition to his scientific qualifications, had been an important consideration in his selection as reader. As the queen wrote to a friend, "he is a philosopher of the proper type because he does not reject religion and since all of his researches are filled with admiration for the Supreme Being who furnishes us with worthy objects to enlighten our hopes; is one able to properly understand nature without admiring the Creator? I think not!"

The position of reader was not easy, and de Luc reportedly spent three or four hours a day standing at a desk in a corner of the queen's chamber, reading scientific works to her as she embroidered or worked tapestry. He also selected scientific books for her to read and assisted with her other scientific interests, particularly botany. De Luc and his wife lived in a cottage just outside the walls of Windsor Castle, near the royal family's quarters, and he soon became a familiar and influential member of the family's inner circle. The trust and respect that the royal family felt for him can be seen in the fact that in 1788, during King George's "madness," his son named de Luc as one of only four people to have free access to him.

In 1779 de Luc published an immense six-volume study that sought to unite science and the Christian Enlightenment themes into a single, unified work. Published in French, the English translation of the title was *Letters Philosophical and Moral Concerning the History of the Earth and of Man*. At more than 3,400 pages and without a single illustration, this extremely dense work was nothing less than an attempt to draft a grand synthesis of contemporary science, scripture, and political philosophy. Significantly, it was dedicated to the queen, which signaled her approval of it.

In this work, de Luc proposed a geological theory in which Earth's history consisted of two distinct phases. On one side was the antediluvian Earth, the period during which Earth had been formed and originally populated. On the other side was the modern Earth, the world of human history. The Deluge, the flood of Moses described in the book of Genesis, separated it from the antediluvian Earth. Based on geological observations that he had made over the course of several decades, de Luc presented scientific evidence of not only the Deluge but also a subsequent series of events from which he measured the age of the modern world to be only a few thousand years old—a measurement that matched the age of human civilization. The goal of de Luc's theory had been to present clear physical evidence that, as he said, "Genesis, the first of our sacred books, contains a true history of the world," and in this he claimed success. The Bible, he asserted, was anything but the "fable" that the deists and atheists claimed it to be.

The idea of the Mosaic Flood as a major geological event was not original to de Luc, but he was the first to present a plausible scientific argument to support it. Even though it was long and difficult to follow, the initial reaction to de Luc's book was extremely positive. The influential *Monthly Review* lavished praise: "We have not, in many years, met with a work more replete with rational entertainment and solid instruction, and which we can more conscientiously recommend to the friends, and also to the enemies, of *true* philosophy, than the work now before us." De Luc's work was also widely accepted by clerics and members of England's upper classes, who saw it as an important bulwark of "the defense of Christian philosophy against Deism and atheism."

In de Luc's work, the Christian Enlightenment ideal of science confirming the Bible seemed to have been achieved, and the institutions that relied on that authority, like the church and the monarchy, placed on a secure foundation. Science, it seemed, really would confirm religion, and the tolerance that the English had traditionally shown toward free and open scientific discourse seemed wonderfully justified. The 1780s, the decade in which Smithson attended Oxford and began his scientific career, was a golden time for English science. Scientific speculation and radicalism—even toward religion—were benevolently tolerated. But with the sudden onset of the French Revolution that began to change. England's governing class was increasingly afraid that a similar upheaval would take place in their own country, and that fear expressed itself in a growing intolerance of anything that seemed to challenge the established order. This new conservatism did not target science per se, but the idea that science should have limits soon gained wide acceptance, and throughout the 1790s, the English grew increasingly suspicious of new scientific ideas.

De Luc played a part in this change. After the publication of his *Lettres* he had moved on to other scientific topics, but in the 1790s he returned to geology and began to sound a shrill new tone about the dangers of ideas that seemed to challenge his system. He began to refer to his main opponents, the geologist James Hutton and those who supported him, as "infidels" and promoted a conspiracy theory in which deist natural philosophers had used "false" geology to destroy the public's faith in the Bible. While these ideas appear to have attracted few actual adherents, de Luc's connections with the king, the Anglican Church, and the upper levels of government gave him a powerful voice and for the most part kept his accusations from being openly opposed.

In this tense environment, mainstream English science as represented by the Royal Society could hardly remain unaffected. The Royal Society was England's most prestigious scientific body, and Joseph Banks, its president, attempted to safeguard it from this new conservatism in several ways, the first being to avoid electing new members who might attract criticism. After 1790 it became increasingly difficult for anyone known to have made radical statements to become a member. By 1796, Thomas Beddoes could complain about "the recent practice of electing Members of the Royal Society from the colour of their political opinions," and by 1804 Banks could proudly write that "we shall not now I trust goe [sic] astray as I think we have not one attending member who is at all addicted to [radical] Politicks."

Banks knew the greatest danger to the Royal Society lay in scientific speculation, especially about geology. As one historian observed, "Conservative hysteria in Britain against the Revolution was quickly to blackball all speculative natural philosophy, all science derived from Enlightenment naturalism, all views of Earth history which seemed to assail Christianity." Accordingly, after the start of the French Revolution there was a broad movement within the Royal Society, and in English science in general, to eschew speculation and concentrate on the collection of facts. This was especially true in geology. In the early 1790s, Royal Society member John Hunter was warned by a colleague against advocating an old-earth cosmology (defined as any estimate of the earth's age longer than six thousand years) for fear of offending the public's "pardonable superstitions." In 1803, when William Richardson, a famously combative clergyman-geologist, submitted a memoir on geology, Banks himself suggested that he remove any discussion of geological theory from it—which he did.

By the early nineteenth century, as fears about revolution began to subside, normal scientific dialog was beginning to resume in England, but not about geology. This is clearly seen in the founding of the Geological Society of London (1807), where it was explicitly stated that debate about geological theories was not part of the society's agenda. Not that all discussion of "philosophical" geology suddenly ceased, as we saw with Humphry Davy's theory. But the debate was largely confined to the halls of Davy's Royal Institution, and during the first two decades of the nineteenth century very little of that discussion appeared in print.

At least part of the reason for this was the lingering belief among the educated classes—and even within the English scientific community—in the Christian

Enlightenment ideal of a correspondence between science and the biblical nar-rative. Specifically, there continued to be a deep and widespread belief that the Mosaic Flood had been a real event, that it was provable by science, that it had occurred about four or five thousand years ago, and that this was the age of human civilization. Far less common on the Continent, this belief was an important char-acteristic of British geology (and British science) throughout this period—and throughout Smithson's lifetime.

To give just a few examples, in his introduction to Georges Cuvier's influen-tial *Essay on the Theory of the Earth* (1813), the Scottish geologist Robert Jameson was pleased to note that the "Mosaic account . . . coincide[d] with the various phenom-ena observable in the mineral kingdom." In his *Critical Examination of the First Prin-ciples of Geology* (1819), George Greenough, a founder and the first president of the Geological Society of London, revealed that he believed in both the Deluge and the associated chronology; and in *Outlines of the Geology of England and Wales* (1822), one of the standard texts of the period, the authors assumed without question that there was clear geological evidence for the Mosaic Flood. The reality of the Mosaic Flood also enjoyed strong support within Oxford University, as seen in the fact that in 1818, the first man appointed to the university's new chair of geology was Wil-liam Buckland, a firm supporter of the reality of the Deluge.

Given this situation, Davy's decision not to publish his theory makes sense. In addition to providing an explanation for how volcanoes worked, the idea that Earth has a reactive metallic core implied that volcanoes were part of an ongoing global process. The existence of this metallic core also provided a new mechanism for geologic change, specifically the "subterranean heat" required by the theory of James Hutton. All of this was scientifically defensible, but it left no room for any connection of geology with the biblical narrative. It was Davy's right to talk about his theory within the confines of the Royal Institution, and it was given exposure in the newspapers and by word of mouth. Had he chosen to publish it as a scientific article, however, he could have become involved in a political and religious argu-ment that he clearly thought prudent to avoid.

Davy's scientific colleagues seem to have come to the same conclusion and stu-diously avoided topics related to geology, and particularly volcanoes. In the first two decades of the nineteenth century, *Philosophical Transactions* published only two articles in which volcanoes were even mentioned. One was a brief report by a Cap-tain Tillard of the Royal Navy, who in 1812 witnessed the creation of a new island as a volcano rose out of the sea. The other was the article by Smithson.

This, then, was the environment in which Smithson wrote his article, and seen from this perspective it now begins to make sense. Both he and Davy were members of the Royal Society and of the Royal Institution's chemistry committee, so he was familiar with Davy's theory. And given Smithson's long-standing interest in volcanoes and his support of Hutton's "vulcanism," it is no surprise that he found Davy's theory appealing. But as a member of the Council, the Royal Society's governing body, he would have also been aware of the concerns about discussing geology. It must have pained him to see science suppressed in this way. Unlike his colleagues, who chose for the moment to remain silent, Smithson chose to speak.

He clearly wanted to write an article describing evidence he had found that supported Davy's theory, but, as we have seen, that theory had yet to appear in the scientific literature. So Smithson's first order of business was to introduce it, which is why the opening of his article is all about geology. What he presented was a summary of Davy's full theory that the Earth had once been a fiery body, that a crust had formed over it to make it habitable, but that underneath that crust was still a hot metallic ball that produces volcanoes when parts of it are exposed to water and air. This was what Davy demonstrated with his final model, and now Smithson had put it in print. In doing so he was careful not to mention Davy's name or to attribute any part of the theory to him. Smithson's article actually reads as though these were all his ideas, although most of his readers would have understood and seen through this protective fiction. The fiction gets a bit thin at the end, though, when Smithson feels the need to present some evidence to support his assertions. Without citing Davy, the best he can offer is an observation from mineralogy: that Earth's oldest rocks fail to show any signs of having been exposed to water.

Having stated this, the tone of the article suddenly shifts, and Smithson's other purpose now becomes clear. If Davy's theory was correct, the material coming out of volcanos should show evidence of the "combustible material" deep underground, particularly the reactive metals potassium and sodium. This is just what Smithson had found in the sample from Mount Vesuvius. He reported finding large amounts of both potassium and sodium. He also found small amounts of the metals copper and iron, all of which seemed to strongly support Davy's theory of the existence of a metallic core. And this is why Smithson didn't go any further with his analysis, even though he suspected there was more to be found: this article was about confirming Davy's theory, and finding these metals had done so.

Not surprisingly, Smithson's article generated considerable scientific interest and was quickly reprinted in the *Philosophical Magazine*. At this point in his career, Smithson was one of England's foremost analytical chemists, and there is no evidence that anyone questioned his analysis. Nor was his suggestion that Earth had originally been a sun or comet considered particularly objectionable—this sort of speculation had long been common when writing about geology, and Smithson

Consequently the soluble portion of the present Vesuvian salt consists of

Sulphate of potash	-	-	7,14
Sulphate of soda	-	-	1,86
Muriate of soda	-	-	0,46
Muriate of ammonia			
Muriate of copper	}	- -	0,54
Muriate of iron			

10,00

Smithson's report on the materials he found in his analysis of the substance from Mount Vesuvius. By weight, nearly 95 percent of the compounds he found in the sample contained either the metals "potash" (potassium) or "soda" (sodium). This was a fact that only Davy's theory could explain. Courtesy of the Smithsonian Libraries.

was known to have a "philosophical" bent. Most of the reaction to his article followed the tone of the summary in *The Monthly Review*, which simply reported on it without comment. *The Eclectic Review* took the same approach, adding only the observation that "Mr. Smithson is a decided Huttonian."

Smithson's paper was read to the Royal Society in July 1813 and published in the *Transactions* later that year. But just a few months later an article strongly critical of it—and of Smithson himself—appeared in the *Journal de Physique*, and, at exactly the same time, an English translation of this article appeared in the *Philosophical Magazine*. The author was the venerable, and still formidable, Jean André de Luc, now nearly eighty-six years old, who continued to fiercely defend the correspondence between science and the Bible. This self-appointed defender of the Christian Enlightenment had clearly seen Smithson's article as a threat and had rushed his own articles into print, determined to refute it in any way he could.

Smithson's paper was just six pages long, but de Luc's reply ran to twelve. Despite its length, it did not directly address anything that Smithson said. The old warrior was clever, and the last thing he wanted was to legitimize Davy's theory by talking about it. So, since Smithson had not mentioned Davy, neither did he. Instead, he played along with the fiction that this was all Smithson's idea.

Smithson's paper, he wrote, is "intended to support a new geological system, thus introduced at the beginning of the paper," and he named that system "Smithson's Hypothesis." "This surely is as new, as it is a grand system," he continued, "but what is its foundation? The author tells it himself in the title of his paper; it is on *a saline substance from Mount Vesuvius*. . . . He then proceeds to detail the chemical analysis which he had made of that *saline substance*; but as he is aware that so small and even so indirect a fact could not support his *igneous system*, he ventures on another ground, in which I shall now follow him." By interpreting Smithson's article in the narrowest possible sense, and then carefully selecting which parts to respond to, de Luc was able to argue that the only real evidence for Smithson's "hypothesis" was his observation that no crystal embedded in the oldest known layer of rocks "has ever been seen enclosing drops of water."

This was a reference to the fact that many kinds of natural crystals can be found with small spaces inside, in which a tiny amount of water can be seen. Today these spaces are called "liquid inclusions," and we believe they occur because the different parts of a crystal "grow" at different rates. A crystal's surfaces, edges, and corners grow the fastest, and can sometimes literally grow over undeveloped interior spaces, creating small chambers and trapping inside them some of the solution from which the crystal formed. Not all crystals form by precipitating from a solution, however, and so the presence of water in crystals was a good indicator of whether they (and the rock layer in which they were found) had formed in a liquid environment. But if this was the case, Smithson argued, it must also be true that if none of the crystals in a given layer of rock contain water, then that layer must have formed in the absence of a liquid. If that layer had been part of the initial "crust" that smothered the burning Earth, this is exactly what one would expect to find.

Smithson described finding crystals of quartz, garnet, and tourmaline in the oldest rocks, and reported that neither they, nor the rock level in which they had formed "contain water in any state." But an argument based on a lack of evidence is always vulnerable, and de Luc took full advantage of this opening. "Now, since the author says in favour of his *igneous system*, that *no crystal such as quartz has ever been seen inclosing drops of water*," he wrote, "which ought to be the case in that system; the contrary being the *fact*, is a peremptory proof against it." De Luc was arguing that if he could produce a single quartz crystal with water inside, it would negate Smithson's system—and he reported having several such crystals in his personal collection.

The existence of quartz crystals with embedded water was already well known among mineralogists and would have come as no surprise to Smithson. Nor is it likely that he would have accepted de Luc's contention that they falsified his "hypothesis." The theory he supported was based on much more than this. But Smithson had chosen not to include that information in his article, and in his rebuttal de Luc pressed his advantage. He characterized Smithson as a naïve amateur who had built a theory on the basis of dubious information, calling him "a very inaccurate observer" with "no real knowledge of the various ejections of

Mount Vesuvius." De Luc referred to the wealth of information in his own books, "with which the author appears to be totally unacquainted."

De Luc's argument was both dismissive and insulting, and it is unlikely that it persuaded many of the members of the Royal Society or many of Smithson's European colleagues. De Luc's goal was not to convert his critics, though, but to silence them and prevent the dissemination of ideas that might (in his view) undermine religious faith. At least in this respect, it appears that he was successful.

If Smithson hoped that his article would rouse his colleagues into action and initiate a scientific dialog, he must have been bitterly disappointed. The English scientific community failed to rise to his defense, and even Humphry Davy continued to remain mute on geological topics. In Europe, where de Luc's criticism was widely reported and Davy's theory was mostly unknown, there was also little reaction. There was some interest in Smithson's suggestion that Earth had once been a comet, but that idea was mistakenly credited to the English chemist Smithson Tennant. For the most part, it was as if the article had never been written.

Were there consequences for Smithson having raised this subject in print? On this point the historical record is frustratingly mute. The few remaining facts are that Smithson's "saline substance" article was published in the fall of 1813, that de Luc's highly critical response to it appeared in January 1814, and that in May 1814, just weeks after Napoleon's abdication, Smithson suddenly moved to Paris, where he would live for most of the rest of his life. He would return to England from time to time in the following years, but it would never again be his home.

Knowing that Joseph Banks was deeply concerned about protecting the Royal Society from becoming entangled in politics and that he had gone to considerable lengths to steer it away from geological questions, why then did Banks make an exception for Smithson and allow his article to be published? The answer seems to be that he did not know about it, that Smithson found a way around him. Royal Society records indicate that Joseph Banks was not present—as he usually was—at the meeting where Smithson's article was read to the society, nor did he preside— as he usually did—over the Committee on Papers meeting where that article was approved for publication. Society records show that George Douglas, the Earl of Morton, presided over those two meetings. As the vice president of the Royal Society, Douglas frequently served in Banks's place if he was unavailable.

If Smithson did go around Banks to get his article published, the venerable president could not have been happy, particularly when he read de Luc's article attacking it. De Luc still had powerful connections, especially within the monarchy, and Banks had recently gone out of his way to avoid antagonizing him. Banks and others may have felt that Smithson had endangered the Royal Society, and the timing of Smithson's move to Paris is suggestive. But whether conservatives forced Smithson to leave England, or his colleagues encouraged him to do so for the good of the society, or he simply decided to leave on his own, we do not know.

In the 1820s, Smithson's article finally received some positive attention. In 1825, Alexander Crichton made Smithson's "hypothesis" a central part of his explanation of why the Earth appeared to have been warmer in ancient times than it was now. "This hypothesis concerning the cause of the central heat, was first started, as far as my reading goes, by James Smithson," he wrote, apparently unaware of the connection to Davy. He marveled that Smithson "seems to have been satisfied with merely throwing out the idea, and to have totally abandoned its development." Three years later M. L. Cordier, while not attributing it to Smithson, used his "hypothesis" to explain the warmth of Earth's core: "There is, therefore, every reason to believe, that the internal mass of the globe is still possessed of its original fluidity, and that the earth is a cooled star, which has been extinguished only at its surface."

However, this revival of interest was short-lived. In 1828 Humphry Davy finally published a detailed description of his geological theory, but by that time he had already abandoned it. He reported that his extensive investigation of Vesuvius had failed to confirm his predictions, and that he would no longer pursue that line of inquiry. Without Davy's theory to support, Smithson's article served little purpose, and it faded into obscurity.

———————+———————

10.

THE COLOURING MATTERS
OF SOME VEGETABLES

THE NEXT ARTICLE Smithson submitted to the Royal Society was read on December 18, 1817, and published the following spring. Nearly five years had passed since the publication of his last article, which had not been well received. Smithson sent this one from France, where he was now living, and its tone suggests that he was preparing for a new phase of his life. "I began a great many years ago," he wrote, "some researches on the colouring matters of vegetables. I have now no idea of pursuing the subject. In destroying lately the memorandums of the experiments which had been made, a few scattered facts were met with which seemed deserving of being preserved. They are here offered, in hopes that they will induce some other person to give extension to an investigation interesting to chemistry and to the art of dying." Accordingly, the article was titled simply "A few Facts relative to the Colouring Matters of Some Vegetables."

Although he did not describe them, the "colouring matters" that Smithson studied would have been familiar to his colleagues in the Royal Society. These were the important "indicators" used by chemists to identify acidic and alkaline substances. Extracted from specific plants such as violets, red cabbage, logwood, turnsole, and turmeric, these liquid "matters" would change color to show whether the substance being tested was acidic or alkaline. For instance, the blue liquid extracted from red cabbage by chopping it up and soaking it in hot water would turn red when an acid was added and green when it was combined with an alkali. If the substance was neither, the color would not change. Some of the matters extracted from other plants, such as the blue liquid from violet flowers, exhibited similar color changes, but both the color of the matter and its specific reaction could vary widely from plant to plant. For example, the yellowish liquid *lignum nephriticum*, extracted from the wood of the "kidney wood" tree (*Eysenhardtia polystachya*), lost its color when an acid was added, but the color could be restored by the addition of an alkali. By comparison, the extract of turmeric root was bright yellow and showed no change when exposed to an acid, but it turned a brownish red when put in contact with an alkali. In the early twentieth century, these materials would all be replaced by "litmus papers"—highly engineered combinations of "indicator" materials that would change color in close correspondence to the pH scale. But these papers did not exist in Smithson's time, so knowing how to prepare and use "colouring matters" was an important part of being a chemist.

The use of these indicators dates back at least to the seventeenth century, and in 1664 the English chemist Robert Boyle made the first organized study of them. Boyle called particular attention to syrup of violet, which was made by combining the juice of violet flowers with water and sugar. The resulting thick, blue liquid had long been used as both a household medicine and refreshing beverage, and some alchemists had noted that sulfuric acid turned it red. But it was Boyle who first noted that all acids turned it red and all alkalis turned it green. He called attention to many other plant materials with similar sensitivities and developed them into simple, reliable chemical tests for determining whether a substance was acid or alkali.

By Smithson's time, a wide range of these natural materials had been identified and developed for specific purposes. They were an important part of a chemist's toolkit, and there was a lot to know. Frederick Accum's *Practical Essay on Chemical Re-Agents, or Tests* (1817) devoted the first fifty pages to making and using just five of these vegetable indicators. Smithson himself used indicator colors throughout his career, including in his first article, where he wrote that "some Tabasheer, reduced to fine powder, was boiled for a considerable time in infusions of turnsole, of logwood, and of dried red cabbage, but produced not the least change [of color] in any one of them." One of the sources he probably consulted in planning these tests was Swedish chemist Torbern Bergman's *Physical and Chemical Essays* (1788). Bergman used a wide range of vegetable indicators in his work, and he made it clear that chemists needed to understand their differences in order to use them intelligently. But in spite of his empirical skill, Bergman failed to offer an explanation of what caused those differences. These natural materials were among the best tools that chemists had, and they reliably defined whether a matter was acid or alkali, yet the mechanism by which they worked was a mystery.

This was the problem that Smithson had set out to solve many years earlier, and had he succeeded he would have put chemistry on a more rational footing. It was an important project, and over time he had systematically tested dozens of different flowers, berries, and other sensitive materials, looking for clues about how they worked. In his article he mentioned the need to use fresh materials, which limited him to working only during those seasons when they were available, and he described "the great delicacy of the experiments, and the great care required in them." But ultimately he was forced to admit in his article that "very little was done." The chemistry of Smithson's time was ill equipped to study the tissues and products of living organisms.

In his article, Smithson mentions studying the coloring matters of twenty-four plants (and two insects), but it is apparent that he tested many more. His approach was to extract the specific "colouring matter" of each type of plant and then test it with various acids and alkalis, noting the results and looking for similarities with the matters of other plants. In his article he identified seven separate categories of matters: turnsol, violet, sugar-loaf paper, black mulberry, corn poppy, sap green, and animal greens. He then described each category separately.

The first one he discussed was turnsol, now called turnsole, a blue-colored dye made from the turnsole plant (*Crozophora tinctoria*). In the Middle Ages it was known as "folium" and used primarily as a pigment for manuscript illuminators. The immature seeds of the turnsole plant were ground to produce a dark green liquid that almost immediately began to turn purple when exposed to the air. To stabilize the color, cloth rags were used to soak up the juice and then treated with a weak alkali, originally urine but later probably lime water. The rags were soaked with juice and treated this way several times until they were completely saturated. They were then dried and sold as dye. These rags could be soaked in water and the resulting liquid treated with different acids and alkalis to produce a range of colors, from blue to purple to red, although not all of these colors were stable. Because the turnsole plant is low-growing and the seeds small, producing large amounts of the dye was difficult. A variety of other materials were sold with the same name, including dyes made from elderberries, mulberries, bilberries, and *Centaureas*, a kind of cornflower. In Smithson's time most turnsole was manufactured in Holland in great secrecy. While still called turnsole, it was sold in dried cakes and much larger quantities. In 1799, it was reliably reported that this turnsole was being made from *Rocella orchil*, a lichen; however, Smithson does not seem to have been aware of this.

Chemists in Smithson's time used these turnsole cakes to make the blue solutions they used in their tests, and Smithson's analysis began with the French chemist Antoine Fourcroy's report that "turnsol is essentially of a red colour." This referred to the color of turnsole's matter when not combined with either an acid or a base, and it was an important thing to know before using it to test an unknown substance. Fourcroy believed that the way the turnsole cakes were manufactured left them with an excess of alkali that had to be neutralized before the liquid could be used. Smithson accordingly tested a solution of blue turnsole with several acidic compounds to see if they would form a precipitate, which they did not. He found that adding an acid to the solution turned it red, while adding an alkali turned it back to blue, which he interpreted as meaning "that the natural colour of turnsol is not red, but blue." It was an important finding. Smithson also made a tincture, dissolving some of the turnsole in alcohol, which he then dried and burned to reveal the presence of potassium. This was what he had done earlier in his experiments on ulmin, and it raised the question, "is the colouring matter of turnsol a compound, analogous to ulmin, of a vegetable principle and potash [potassium]?" It seemed likely, but this was as far as early nineteenth-century chemistry could go.

The next set of tests were on "the colouring matter of the violet," and here Smithson was looking at a material that went all the way back to Robert Boyle. "The violet is well known," he wrote, "to be coloured by a blue matter which acids change to red; and alkalies and their carbonates first to green and then to yellow." He found that other plants seemed to have this same matter, too—red rose petals, for instance, had it, as well as some acid that affected the blue matter and made it red. If water with a mild alkali was added to ground up rose petals, a blue solution with all the properties of the matter of violets would be produced. Smithson, it seemed, had found an explanation for the colors of dozens of plants. "The colouring matter of the violet," he wrote, "exists in the petals of red clover, the red tips of those of the common daisy of the fields, of the blue hyacinth, the holly hock, lavender, in the inner leaves of the artichoke, and in numerous other flowers. It likewise, made red by an acid, colours the skin of several plumbs [sic], and I think, of the scarlet geranium, and of the pomegranate tree. The red cabbage and the rind of the long radish are also coloured by this principle." Smithson's suggestion that these plants contained not only the coloring matter but also an acid, provided one of the first suggestions as to how the colors of flowers and fruits could change as they ripened.

Smithson had made an important discovery, a new vegetable "principle," and as the discoverer it fell to him to name it. In his original manuscript, still preserved in the Royal Society's archive, Smithson wrote, "As this colouring matter is a distinct vegetable principle it requires a name. That of Ajax, whose blood is fabled to have died [*sic*] the violet, seems naturally to furnish it." But the suggestion was deleted by the editor and never published.

The third category Smithson described was "sugar-loaf paper." Until the late nineteenth century, refined sugar was sold in the form of a hard, tall cone with a rounded top, commonly called a "sugar-loaf." These came wrapped in blue paper that enhanced their whiteness. Smithson reported that Bergman used this paper in some of his tests, although Bergman admitted, "I am ignorant of what it is coloured with." It now appears that the dye used to color this paper was made from lichens, of the type from which litmus is extracted, and it was a reliable indicator of acidity. Smithson found that, like turnsole, it did not react to alkalis, but that strong acids made it turn red. His tests seemed to show that the paper contained at least two coloring matters: one that was water-soluble, and one that could be extracted only by an acid. He confirmed that neither was the coloring matter of turnsole or indigo. But he seems to have been unimpressed with the usefulness of sugar-loaf paper to chemists, and he did not investigate it further.

The fourth category was "black mulberry," which produced a fine red-colored juice. Mild alkalis made it blue, and strong alkalis made it turn green. Acids turned it a "vinous" red. When Smithson passed the liquid through a filter, it produced a red liquid and left a blue material in the filter. Smithson reported that "caustic potash instantly made this red liquor a fine green, and gradually yellow," while sulfuric acid turned it a "florid red." The material caught in the filter was unaffected by caustic potash, but turned red when sulfuric acid was poured on it, and when caustic potash was put on those red areas, it turned them back to blue. From this Smithson concluded that there were two separate coloring "principles" in the mulberry juice, and he proposed for them the names Pyramus and Thisbe, after the star-crossed lovers from Greek mythology who met beneath a mulberry tree. The editor again chose not to publish Smithson's suggestion.

The next group Smithson described was the corn poppy, "the common red poppy of the fields." He started by simply rubbing the petals on a piece of paper, which left a reddish-purple stain. A solution of strong alkali turned that stain green, but acids seemed to have little effect. When he mashed some of the petals in a bowl with water and hydrochloric acid, it produced a "florid red" solution that mild alkalis turned to "a dark red colour exactly like port wine." Mashing poppy petals in a bowl with water and an alkali also produced a port wine–colored solution, which turned green when he first added a strong alkali, but then slowly changed to yellow. "These very imperfect experiments," he concluded, "may perhaps suggest the idea, that the colouring matter of this flower is the same as the red colouring matter of the mulberry."

This page from Smithson's manuscript suggests that the name "Ajax" be used for the vegetable "principle" of violets. The "X" through the paragraph was made by the editor of the Royal Society's journal, telling the printer not to include it. ©The Royal Society.

The sixth group of matters was sap green, "the inspissated [thickened] juice of the ripe, or semi-ripe, berries of the buckthorn." In Smithson's time this was known in France as *vert de vessie* and used as a water-color paint pigment. Because it was water-soluble, it was convenient to use this pigment to make chemical test papers, and Smithson reported that it was a sensitive test of alkalis, which turned it from green to yellow. Acids, conversely, made it red. Smithson identified this as yet another new "principle," and in his manuscript he suggested the name Chloris, another reference to Greek mythology. Again, the editor chose not to publish his suggestion.

Almost as an aside, Smithson ended his article with a few sentences about what he called "animal greens." He mentioned a juice produced when an insect called "puceron" (or "aphis") is crushed, leaving a green stain that was turned yellow by alkalis, but he seems to have included this mainly to identify it as being different from sap green.

This concluded Smithson's article. After years of work on this topic, he had failed to provide an explanation for how the indicator colors worked. Still, he had identified four new materials, described many more, and suggested a mechanism for how the colors of flowers and fruits change over time. It was an important paper, and he must have been shocked when he read the version printed in the *Philosophical Transactions* and realized that some of the most important parts had been deleted.

Considering both the importance of the topic and the fact that Smithson had been a member of the Royal Society's governing council, it is surprising that these omissions occurred. But in this period authors had few rights once a manuscript left their hands. It should be noted that Smithson's choice of names for these new materials flew in the face of the reformed chemical nomenclature that was being adopted at the time. Terms like those that Smithson proposed, with their references to obscure Greek mythology, were difficult to learn and remember, and were increasingly seen as creating a barrier to scientific progress. However, in deleting those names, the editor also deleted the fact that Smithson had identified those materials as "principles," and when Smithson's article appeared in print it appeared to many to be just a collection of facts with no real point.

When Smithson's article was read to the Royal Society, it was read as he had written it, and the overall reception was favorable. But when the article was published, there was almost universal agreement that it was incomplete. Both the *Monthly Review* and the *Annals of Philosophy* found it disjointed and lacking a conclusion. Smithson's article was also noted in a few foreign journals, and it was reprinted in the *Repertory of Arts, Manufactures, and Agriculture*, which frequently reprinted scientific articles related to agriculture, but otherwise it failed to attract much interest. What did attract interest was Smithson's suggestion that certain plant colors, like the red in rose petals and the red in fruit as it ripens, are the result of their blue coloring matter being combined with an acid. Brande, the chemistry lecturer at the Royal Institution, mentioned this in his textbook.

Today, we call the materials that color vegetables and fruits "chromophores." Many different chromophores occur among the various molecules of plants, but the colors of fruits and vegetables fall into four distinct categories of pigments: chlorophylls, carotenoids, anthocyanins, and anthoanthins. Smithson's research appears to have touched upon several of these categories, but most of the coloring matters that he studied were anthocyanins. The term comes from the Greek words *anthos* ("flower") and *kyaneos* ("dark blue"). Anthocyanins are water-soluble, and they exist in various tissues of higher plants, including flowers, fruits, seed coats, leaves, stems, tubers, and roots. They exhibit a variety of colors, ranging from red to purple and from blue to black, and by 2009 nearly one thousand such pigments had been reported. This is still an active area of investigation, and progress has come through advances in structural determination, as well to improvements in purification methods.

What Smithson clearly hoped to find, but had no way of ascertaining, was the mechanism by which acids and alkalis change the color of these materials. The modern understanding is that changing the pH changes the physical structure of anthocyanins, which in turn affects their absorption of light and thus alters their color. It is also interesting to note that modern research has shown that the anthocyanins in violet flowers are stabilized by joining with sugar—offering a mechanism for the observation from Robert Boyle's time that adding sugar to the juice of violet flowers (to make syrup of violet) helps to preserve it.

Almost a full century after Smithson's article, in 1913, Richard Willstätter and Arthur Everest wrote a classic paper on flower pigments. For the first time, they used pH to explain how flower colors develop. Their hypothesis was based on the observation that cyanidin, a blue pigment from cornflowers, is also found in red rose petals. Their tests in aqueous solutions of various pH values showed that the pigment turns a red color in the presence of acids. Thus, they proposed that the pH differences in the petal cells of blue cornflowers and red roses cause the differences in their petal colors. These researchers were clearly not familiar with Smithson's article, but the similarity between their finding and his suggestion of the same mechanism is striking. Smithson's anticipation was later tacitly acknowledged when his article was listed in the bibliography of "Chemistry of Anthocyanins" (1916).

Considering the effort and years of research that he expended on it, the treatment Smithson's article received from the Royal Society must have been a great disappointment to him. He is known to have become estranged from the organization at about this time, and the cause of the rift has long been an open question. Around 1870 the physicist Charles Wheatstone provided a clue: he reported that in the late 1810s Smithson "became offended with the [Royal Society] Council for having stricken out some sentences from a communication which he presented." A slightly different version of this story had come a few years earlier from Harvard professor Louis Agassiz. Citing information "which I happen accidentally to know," he reported that "Smithson had already made his will and left his fortune to the Royal Society of London, when certain scientific papers were offered to that body for publication. They were refused; upon which he changed his will and made his bequest to the United States."

Previous Smithson biographers have been reluctant to attribute his break with the Royal Society to this article, believing that it was too minor a work to deserve such a strong reaction. Yet, had it been published in full, it seems likely that it would have been widely read and seen as an important contribution. As we saw with Smithson's work on ulmin, the discovery of new plant "principles" was a topic of great interest at this time, and this article could have established him as a leading figure in plant chemistry. Instead, this became the last article he would ever submit to the Royal Society.

11.

A SULPHURET OF LEAD
AND ARSENIC, AND
"PLOMB GOMME"

BY 1819, when Smithson's next article was published, he was in his mid-fifties and living in Paris. Although he would continue to be a member of the Royal Society, he would no longer send papers to be read at its meetings and published in its journal. From now on his articles would go to his distinguished friend Thomas Thomson, the editor of *Annals of Philosophy*. The articles themselves would also change. Where they had previously averaged eight pages in length, the articles he now wrote averaged only three. There was a qualitative difference as well. The articles he had written for the Royal Society had always been noteworthy for their exploration of the philosophic implications of his findings, whereas the articles he would send to Thomson tended simply to report interesting discoveries. Most were noteworthy for their *lack* of speculation, and there is frequently a sense that Smithson was clearing out his files—reporting on topics that he had no intention of revisiting, but that might be of interest to others.

This can be seen in the first paper he sent to the *Annals*, which was a report of a new mineral. Barely two pages long and titled simply "On a Native Compound of Sulphuret of Lead and Arsenic," it described a mineral he had discovered many years earlier on a trip to Switzerland. He described the mineral as being gray with a "metallic aspect." He also noted it was brittle and soft, with a "tabular" crystalline form, but because he only had a few fragments and the largest of these were barely bigger than a particle of coarse sand, this was as much as he could say.

Smithson's sample had come from the eastern part of Switzerland's Valais Canton, in the remote Binn valley, probably from the old Lengenbach Quarry. That site is still of interest to mineralogists because of the veins of unusual minerals that run through its towering limestone cliffs. It is particularly known for the "sulfosalts" it produces—unusual combinations of sulfur and various metals, many of which are found nowhere else in the world. Smithson probably did not visit the site, which is difficult to reach and could be worked only during the summer months. More likely he purchased samples of the mine's products from a Swiss mineral dealer, and only much later, when he was cleaning and inspecting them, did he discover the tiny, dark specks of metal that he now reported.

Sulfur readily combines with metals, forming minerals that are called "metallic sulphurets." The "sulphuret" that Smithson found was a previously unknown combination of the elements sulfur, lead, and arsenic. In 1808 he had written about one that contained sulfur, lead, antimony, and copper, and in that study he used the mineral's composition to speculate about the nature of chemical combination, but in this new article he made no attempt to generalize. This time Smithson satisfied himself with simply reporting a new finding, albeit one he had made many years earlier.

Smithson was the first to report this new mineral but, probably because his sample was too small to allow him to fully describe it, the discovery attracted little attention. His editor Thomas Thomson pointed out that it was a new mineral species, the American mineralogy professor Parker Cleaveland included it in the next edition of his mineralogy text, and the *Journal de Physique* dutifully posted a brief report on the article, but none of the other European scientific journals followed suit, and Smithson's discovery failed to earn a place in the mineralogical literature. More than a decade later the first "official" notice of the mineral was recorded, and that was in a reference to "binnite," a generic term applied to all of the quarry's gray metallic minerals. Eventually, the name "Dufrenoysite" was given to the mineral Smithson discovered, a name chosen to honor a French mineralogy professor. That name is still used.

Mineralogists may not have noticed Smithson's discovery, but the methods he used in analyzing it proved to be of considerable interest to chemists, and many of them soon adopted these new and very sensitive tests, especially the ones for arsenic and sulfur. Probably because he wanted to save the biggest particles of the mineral for his collection, Smithson notes that the pieces he sacrificed for the analysis were "little more than visible." With such a limited amount of material, he needed to plan his analysis carefully and make the most of every test.

As he often did in these situations, he started by testing the substance with the flame of the blowpipe. Smithson reported that the mineral "melted instantly on contact with the point of the flame" and "forced out a quantity of fluid matter" as it cooled. With further heating he reported that it "occasionally swelled up, and puffs of dense smoke issued from it." Blowing harder into the blowpipe produced a more intense heat, and a second kind of smoke emerged, after which the sample fused into a whitish metal ball as it cooled. The ball proved to be lead, although Smithson did not bother to explain how he knew this. Lead was easy to identify and this would have been an uninteresting detail for most of his readers.

Smithson probably suspected that the new mineral contained both sulfur and arsenic as soon as he started to heat it. Sulfur has a low melting point and readily produces a thick smoke with the distinctively unpleasant smell of rotten eggs. Arsenic does not melt under those conditions, but at higher temperatures it volatilizes to produce a thick smoke with the distinctive smell of garlic. One of the features of the blowpipe is that the user's nose is just inches away from the sample being heated, so Smithson doubtless experienced those smells and used them to guide his analysis. This would explain his next test, which was to place a piece of the mineral in the end of a small glass tube and melt it. As the smoke from the sample collected in the tube and condensed, it formed a substance with a distinctive yellow hue that he described as being the same color as "orpiment"—a yellow mineral known to consist of sulfur and arsenic.

Using the blowpipe to test the mineral, Smithson had shown that it contained lead, and probably sulfur and arsenic as well. He now turned to "wet analysis" to chemically confirm the presence of these last two elements, and the novel process he devised to do this shows why he was one of the most respected analytical chemists of his time. The first step was to isolate the arsenic, which he did by melting it with a small amount of "nitre" (niter, potassium nitrate), and then dissolving the resulting mass in a few drops of water. Adding "muriate of barytes" (barium chloride) to this liquid produced a white solid that settled to the bottom, taking the lead and the sulfur with it. Any arsenic that had been in the mineral was now in the liquid, and Smithson demonstrated its presence by adding a tiny amount of "nitrate of silver" to the liquid, which almost magically produced a distinctive brick-red precipitate that he knew from its color to be "arseniate of silver"—silver arsenate.

Having demonstrated the presence of arsenic, Smithson now returned to the white solid he had produced with the barium chloride. This was where the rest of his sample now was, but before he could test the lump for sulfur he needed to remove the lead. He did this by putting it in a few drops of nitric acid, which dissolved the lead and left behind a quantity of insoluble material that he described as being "so minute that no balance would have been sensible to it." If there had been any sulfur in the sample, this was where it now was. After carefully collecting and placing it on a tiny piece of charcoal that he held on the end of a pin, he heated it with his blowpipe to convert it to a sulfide. Then he performed the final step of the test: "This bit of charcoal now put into a single drop of water, placed on a silver coin, immediately made a black stain of sulphuret of silver on the coin. This is the nicest test I am acquainted with of the presence of sulphur, or sulphuric acid, in bodies."

The stain that Smithson described was the result of sulfur combining with silver to produce silver sulfide, which is black. These two elements combine readily, and the fact that Smithson's test could detect an amount of sulfur that easily fit into a single drop of water gives an indication of its extreme sensitivity. Indeed, the test was even more sensitive than that, since smaller amounts of sulfur would still make a brown stain.

Although he did not take credit for thinking of them or even for being the first to use them, the tests Smithson described in his analysis expanded the capability of early nineteenth-century chemistry. In addition to its use in mineralogy, Smithson's arsenic test soon found applications in both medical and judicial cases, and chemists quickly adopted his sensitive sulfur test. In several of his books, Swedish chemist Jacob Berzelius credited Smithson for introducing the technique, which he described as "both simple and delicate," and "Smithson's Methode" was later developed to detect other substances as well. The use of silver and silver compounds to detect sulfurous materials continued to be an important chemical test well into the twentieth century, and the development of silver staining as a histological technique in the mid-nineteenth century can be seen as a related development. However, by the mid-1830s Smithson's name had already ceased to be associated with this test, and his contribution to chemical precision—just like his discovery of the new sulphuret—was forgotten.

On May 22, 1819, just three days after sending off his article on the sulphuret, Smithson mailed another one to his publisher. This time the topic was a curious mineral called "plomb gomme." A sample had been in his possession for several years, and there was a sad story associated with it. The rare mineral had first been reported in the late eighteenth century, when it had been found in the historic Huelgöet lead mine in northwest France. In Smithson's time this was the only place it was known to occur. The "plomb" part of its name referred to the fact that it contained lead, and "gomme" to its resemblance to the resinous gum sometimes exuded from trees. The first scientific notice of it came in 1783 when French crystallographer Romé de l'Isle described it, and in 1786 the French mineralogist Gillet de Laumont analyzed it. Laumont initially reported it as *"sel acide-phosphorique-martial"*—a combination of phosphoric acid and iron—although he later revised that determination. Laumont seems to have taken a particular interest in this unusual mineral, and as France's Inspector of Mines he had ready access to it.

In 1814 Laumont gave a sample to the Cambridge chemist Smithson Tennant, who was visiting Paris, and asked him to analyze it. James Smithson was also in Paris. The two men were longtime friends and distant relatives, and they seem to have spent much time together before Tennant left in early 1815. Indeed, they spent one of Tennant's last nights in Paris gambling together, and the news that Tennant won £100 from Smithson quickly made the rounds among their friends. They also talked about science, and in his article Smithson reveals that "Mr. Tennant mentioned to me a sort of explosion occasioned by the sudden expulsion of the water [from plomb gomme], and characteristic of this ore, which took place when it was heated at the blow-pipe." Smithson, who had also analyzed this mineral, had used much smaller samples and failed to observe this property.

Tennant found that plomb gomme was actually a combination of lead oxide, aluminum, and water. He informed Laumont and Smithson of this, but failed to publish his findings. Shortly after his departure, the shocking news reached Paris that he had died in an accident. Riding near Calais, Tennant had been thrown from his horse, suffered a skull fracture, and died on the spot. He was fifty-four.

Smithson's sadness over the loss of his friend must have resurfaced four years later when he learned that Laumont had given a sample of the strange mineral to Jacob Berzelius, another of Smithson's friends, and that Berzelius had recently published an analysis of it. "I see in the *Annals of Philosophy* for this month," Smithson wrote, "which I have very lately received, an analysis by M. Berzelius of the mineral which was formerly known here under the name of 'plome gomme.' The first discovery of the composition of this singular substance belongs, however, to my illustrious and unfortunate friend, and indeed distant relative, the late Smithson Tennant."

Smithson wrote this in an article titled "On a Native Hydrous Aluminate of Lead or Plomb Gomme." He had no problems with Berzelius's analysis, which agreed with Tennant's, but he was determined to see that his late friend got credit for being the first to make it. As evidence, he quoted a note that Laumont had given him several years earlier, along with a sample of the mineral. The note read "hydrate of alumina and lead recognized by Mr. Tennant, from Huelgoat, near Poullaouen, Brittany."

Smithson's gesture of seeking credit for his friend was both generous and unusual, but he did not dwell on it. The rest of the article consisted of a brief description of Smithson's own analysis of the mineral, which mirrored that of his friend. The fact that it was a hydrate—that it contained water—was demonstrated by placing a small piece of the mineral in one end of a glass tube (a "Berzelius tube") and heating it over a candle. As the sample cracked and began to fall apart, water vapor was released and condensed as tiny droplets in the far end of the tube.

The presence of lead was demonstrated by a blowpipe test. A bit of the mineral was melted with borax and then again with potassium nitrate, which produced small globules of metallic lead. The presence of aluminum was demonstrated by another common blowpipe test, which began with directing the flame on a sample of the mineral until it turned white. When wetted with a few drops of nitrate of cobalt and again put under the flame, it turned blue. This was a common test, and the blue color indicated the presence of aluminum. Smithson concluded, "The above characters will prove sufficient, I apprehend, to make this substance known when met with."

Smithson's article was just one page long, and it repeated an analysis that had already been published by Berzelius, one of the greatest chemists of the age. Nonetheless, it soon became part of the scientific literature. Thomas Thomson, editor of the *Annals*, set the tone by seeming to credit Smithson's article rather than Berzelius's—even though both articles had appeared in his journal, and Berzelius's had been published first. He also carefully noted that "Its nature was first ascertained by Mr. Tennant." An even stronger endorsement came in Sowerby's authoritative *Exotic Mineralogy* (1820), which listed Laumont, Smithson, and Berzelius in its bibliography and noted in the discussion that "it was ascertained by the much to be regretted and unfortunate Smithson Tennant, Esq. when last in Paris, to be a compound of Oxide of Lead, Argilla [Aluminum] and Water."

This understanding remained unchanged until 1835, when Armand Dufrénoy, Laumont's successor, analyzed samples of plomb gomme from a newly found deposit. Surprisingly, he reported that in addition to lead, aluminum, and water, they contained phosphorus, and he suggested that this had been missed by previous investigators. Dufrénoy thought they had been misled by a quirk of the test they used to detect aluminum. Smithson, Tennant, and Berzelius had all used the "nitrate of cobalt" test to detect aluminum. It involved heating the sample with a blowpipe, putting a few drops of nitrate of cobalt on the residue, and heating it again. If the sample turned blue, it contained aluminum.

This test came from the work of the French chemist Louis Jacques Thénard, who had been searching for a new blue pigment and in 1802 synthesized one by combining cobalt and alumina. Cobalt aluminate, or "Thénard's blue" as it was known, was quickly adopted by artists and utilized by chemists. Chemists found that if adding nitrate of cobalt to an unknown substance and heating it turned it the color of Thénard's blue, then it contained aluminum. It was a reliable test, but there was a problem. It seems that adding nitrate of cobalt to phosphorus also produces a blue color, a blue just a bit deeper than the blue produced by aluminum. Dufrénoy suggested that if plomb gomme contained both aluminum and phosphorus—a rare combination—the blue of the phosphorus could have been masked by the blue of the aluminum.

The idea that Tennant, Berzelius, and Smithson—three of the leading chemists of their time—could all have made the same mistake seemed unlikely, but Dufrénoy's chemical analysis was solid and supported by further investigation. Because all three had used samples provided by Laumont from the same mine, it was possible that they tested a form of plomb gomme that simply did not contain phosphorus, but after so many years this was hard to determine. For the rest of the nineteenth century, the composition of plomb gomme remained uncertain. Smithson's and Berzelius's analyses continued to be cited in mineralogical works, and as late as 1914 Edward Dana's influential *Descriptive Mineralogy* still listed the composition of "plumbogummite," as it was now called, as "uncertain." But in the twentieth century, with no examples of plomb gomme having been found without phosphorus, Smithson's and Berzelius's analyses essentially became a scientific footnote, and Tennant's name was no longer mentioned.

Smithson's determination to secure the reputation of his friend Tennant speaks to his values and his character. This story also speaks to the way that science was practiced in Smithson's time, relying heavily on personal relationships and the exchange of favors. Laumont's generosity in providing mineral specimens to his colleagues was typical, as was Tennant's willingness to analyze those specimens without compensation. Smithson himself often provided this same kind of service, as can be seen from a short note that appeared in the *Annals* in 1814. It was reported that specimens of pure aluminum had been found near Brighton, on the south coast of England, and Tennant had arranged for some to be brought to London to be examined. Specimens were given to both Smithson and his Royal Society friend William Wollaston for analysis, and the *Annals* reported that "both have found it to be not pure alumina, but a subsulphate of alumina." All this happened without funding, supervision, or the involvement of schools, institutions, or the government. This was "gentlemen's science," and in Smithson's time it frequently worked.

———————+———————

12.

FIBROUS COPPER AND CAPILLARY METALLIC TIN

MINERAL COLLECTIONS SOMETIMES include "slag," the rocklike waste products left behind by the extraction of metals from their ores. Metal foundries and refineries produce significant amounts of slag, and, while not actually minerals, these materials are often interesting to mineralogists because they bring together substances not commonly combined in nature.

Smithson had slag samples in his mineral collection, and in March 1820, he wrote about one that he had collected many years earlier at an iron foundry in "the Hartz," the Harz Mountains region of what is now central Germany. He had seen examples of this particular slag in other mineral collections, and what interested him about it was that, when broken apart, it was found to contain small holes filled with hairlike copper fibers. His article was titled "On a Fibrous Metallic Copper." Smithson described the fibers as "so delicately slender as to be a metallic wool," and for a long time he had been baffled as to how they could have formed. This form of copper had not been found in nature, and the process by which it could have been made was a mystery. The usual explanation for materials found in the cavities of a rock was that they had crystallized from groundwater, but Smithson wrote that for several reasons "the opinion of these fibers having been produced by crystallization was perfectly inadmissible."

He admitted to having been "for a very long time totally unable to come to any conjecture with respect to the mode in which they had originated," but then the solution suddenly came to him. "Looking on one of these specimens this morning," he wrote, "an idea struck me which is, I am convinced, the solution of this knotty problem." Smithson realized that the fibers must have been extruded. They had, he reasoned, formed at the moment that the molten slag began to harden. As the outer surface of the slag began to cool and shrink, "it had compressed drops of copper, still in a fluid state, dispersed in its substance, and squeezed a portion of it through the minute spaces between its particles, under this fibrous form, into the cavities, or air-holes." He compared the process that formed the fibers "to that employed for the manufactory of macaroni and vermicelli . . . which are made by forcing paste through small apertures by the pressure of a syringe."

Smithson listed the unusual set of conditions he thought needed to exist for the fibers to be created: "The slag must be so thick and pasty as to retain metallic copper scattered through it. It must have developed bubbles of some gas which have occasioned vacuities in it. [And] it must be less fusible than the copper, but in so very small a degree that the copper consolidates as the fibres of it are formed."

To test his theory, Smithson melted a small piece of the slag and, as soon as it cooled, broke it apart. The result was just what he had hoped for. "I had the gratification," he wrote, "of finding its little cavities lined with minute fibres of metallic copper." Having demonstrated the process by which the fibers were formed, the next step was to produce them from raw materials, and for this Smithson turned to his mineral collection. An earlier analysis had shown the slag to consist of copper, iron, and sulfur, so he chose to work with the yellow mineral chalcopyrite, which contains all three elements. It took several attempts before he learned how much heat to apply, but he soon reported having "succeeded in producing a little mass of slag, whose internal cavities presented me, on breaking it, with the fibres of copper which were the object of my toil."

Smithson had discovered a new physical process. "The power," he wrote, "to which has been ascribed the phenomenon which forms the subject of these pages has hitherto been overlooked." It took him just a single day to do it. He got the idea in the morning, spent a few hours on the experiments to confirm it, and then used the rest of the day to write a two-page article describing it. He closed with a few comments about the special conditions required for the process to work and concluded with some thoughts about the importance of understanding how things are made and the properties of materials: "A knowledge of the productions of art, and of its operations, is indispensable to the geologist. Bold is the man who undertakes to assign effects to agents with which he has no acquaintance; which he never has beheld in action; to whose indisputable results he is an utter stranger; who engages in the fabrication of a world alike unskilled in the forces and materials which he employs."

The process that Smithson suggested was ingenious, but his colleagues did not find the pasta-making analogy convincing. The idea that a process like the one he proposed could take place inside the small samples he described using must have seemed extremely unlikely. As a result, unlike any of Smithson's previous articles, this one was greeted by an almost complete silence. Only the *Annaes das Sciencias*, a Portuguese-language journal published in Paris, seems to have noticed it.

Undeterred, Smithson continued to think about the process and talk about it with his colleagues, and he soon learned about another example of pressure producing a metal "wool." The information came from an unexpected source: the French physicist André-Marie Ampère, who told Smithson about a colleague who had made a copper cylinder for some undisclosed purpose. As Smithson described it, "to give it strength, [he] introduced into it a hollow cylinder, or tube, of cast iron. To complete the union of these two cylinders some melted tin was run between them." Ampère's colleague expected the tin to bond the two cylinders together, but "during the cooling of this heated mass, a portion of the melted tin was forced by the alteration of volume of the cylinders through the substance of the cast-iron cylinder, and issued over its internal surface in the state of fibers, which were curled and twisted in various directions."

Just as in the slag that Smithson studied, the pressure of contraction had squeezed the liquid metal through the pores of the material containing it, and the metal had emerged as fibers. The only real difference between the two cases was that Smithson found copper fibers, whereas in Ampère's story the fibers were tin. In both cases they were remarkably thin, curled, and looked like metallic "wool." Smithson quickly wrote an article, "On Some Capillary Metallic Tin," following up on the one he had written about the slag eleven months earlier.

Smithson also alluded to another example of mechanical pressure—the famous "Florentine Experiment." This experiment, first performed in the seventeenth century by Francis Bacon to determine if water could be compressed, was one of the core experiences of Western science. Bacon described having "a leaden [lead] globe made, with very thick sides, and a small hole at the top. This globe I filled with water, and then soldered up the hole (as I remember) with metal. I then forcibly compressed the globe at the two opposite sides, first with hammers and afterwards with a powerful pressing-machine. Now when this flattening had diminished the capacity of the globe by about an eighth part, the water, which had borne so much condensation, would bear no more; the water admitted of no greater condensation; but on being further squeezed and compressed it exuded from many parts of the solid metal, like a small shower."

Bacon's conclusion was that under sufficient pressure, water could pass through the "pores" of solid metal. This experiment was famously repeated in Florence with a silver sphere, and Isaac Newton (along with many others) later repeated it with a gold sphere, but the results were always the same: as pressure built up inside, water would begin to come through the sides of the vessel, eventually covering the outside of it in what looked like a fine, dripping dew.

Smithson made a point of linking his theory about how fibers were formed to this famous experiment. "This passage of melted tin through cast-iron," he wrote, "has a perfect agreement with the passage of water by pressure through gold, and tends to elucidate and confirm the account of the celebrated Florentine experiment. Had the water on that occasion issued solid, it would have been in fibers." Placed in this context, the process he was proposing looked more like an extension of what was already known.

Smithson's strategy worked. This second article was well received and noted in journals of several languages on both sides of the Atlantic. It also elicited two more examples of capillary fibers. "Mr. Smithson's hypothesis, that the capillary copper found in the cavities of copper slags is produced by propulsion, is not less ingenious, than the experiments by which he endeavoured to verify it appear conclusive," wrote Charles Konig, a curator at the British Museum. Konig then went on to describe finding silver fibers "in the hollows of porous iron-shot quartz from Peru," and mentioned that Smithson's friend James Hall had inadvertently produced metal wool in some of his experiments. Several years earlier, Hall made a series of pioneering experiments on the effects of heat and pressure, using iron gun-barrels as the vessels in which the experiments were performed. As Konig described it, "a substance like wool was formed in several of those experiments by the exudation of the fusible metal through the [sides of] barrels of iron employed by him." Konig reported that Hall deposited some of this "wool" at the British Museum and that it was still in its collection.

The phenomena of matter under high pressure were philosophically interesting in the nineteenth century, but not of any immediate, practical importance. In 1879, when the Smithsonian came to evaluate Smithson's achievements, his articles on metallic fibers were praised more for their form than for their content. "Perhaps the most finished of his papers is that 'On a Fibrous Metallic Copper,'" the reviewer wrote, "combining, as it does, an ingenious explanation of a singular phenomenon and subsequent confirmatory experiments. His style, so clear, so direct, and so exact, is a model for scientific purposes."

But Smithson had foreseen that these phenomena, as he put it, "will, perhaps, come to be thought deserving of more attention than has been yet paid," and in the early twentieth century this proved to be the case. Studies of how materials react under the effects of extreme pressure rewrote the scientific understanding of matter, and as that understanding was codified, Smithson was seen as having made an important contribution. In J. W. Mellor's magisterial (and at sixteen volumes, aptly named) *Comprehensive Treatise on Inorganic and Theoretical Chemistry*, is the following brief statement: "J. Smithson found that tin could be forced through cast iron much as water can be forced by press[ure] through gold." It is the kind of terse acknowledgment that Smithson would have appreciated.

———————+———————

13.

SULPHATE OF BARIUM AND FLUORIDE OF CALCIUM

SMITHSON FINISHED THIS ARTICLE on March 24, 1820, seven days after sending off his previous article about "fibrous copper." The title was "An Account of a Native Combination of Sulphate of Barium and Fluoride of Calcium," and it began with him studying a mineral in his collection, one he had discovered many years earlier in Derbyshire, England.

He described finding it in a limestone deposit, where it had formed in a vertical vein about an inch wide. On one side of the vein, next to the limestone, was a thin, distinctive layer of the silvery crystals of "sulphuret of lead" (PbS), and on the other side a thin layer of whitish crystals of "carbonate of calcium" ($CaCO_3$). The rest of the vein was filled with a less distinctive material that he described as having "in its general appearance so strong a resemblance to fine compact grey limestone that the eye can probably not distinguish between them." Smithson did not say what initially drew him to investigate this nondescript material, only that "I have examined several minerals which in appearance bore a resemblance to it, but have not found any of them to be of the same nature. This species would hence appear to be of rare occurrence in the earth."

Without further comment, he then presented a physical description of the mineral. He measured its specific gravity as 3.750 and noted that it did not absorb water. He tested its hardness and reported that it would scratch "sulphate of barium" and that "its hardness and that of fluoride of calcium appeared to be the same." He reported that it did not lose weight when heated, that heat did not produce any electrical charges in it—although the friction of rubbing it did—and that it became transparent when melted but turned opaque as it cooled. When melted with borax, which was a common blowpipe test, the mineral dissolved into a brown glass-like material that had "a fine hyacinth colour." Smithson interpreted this last test as indicating that the mineral contained sulfur.

The presence of sulfur was confirmed by another test. As Smithson described it, a single drop of water was placed on a piece of silver, and a tiny piece of the stone that had previously been melted and cooled was placed in the drop. This immediately made a black spot appear on the silver, underneath the drop of water. This test, which would come to be called "Smithson's method," was a sure indication that the mineral contained sulfur.

The next test involved placing a drop of hydrochloric acid (HCl) on a piece of glass and then placing another bit of the melted stone in the drop. Smithson reported that the stone "partly dissolved with effervescence," and that as the liquid subsequently dried it produced tiny crystals that he identified as barium chloride ($BaCl_2$). The chloride had come from the acid, but the barium must have been in the stone.

In the third test Smithson also used a piece of glass, but this time he placed a drop of sulfuric acid on it. Adding some of the powdered mineral to the drop and heating it gently with the blowpipe produced a distinctive effect. As Smithson described it, the polish of the glass was destroyed in the area underneath the drop. Only fluoric acid was known to produce this effect. This standard test could mean only one thing—that the mineral contained fluorine.

Although this is the order in which Smithson described his experiments, there is no reason to think that this was the order in which he made them—or that he didn't make other experiments that he chose not to mention. But it does seem clear that at some point early in his analysis he suspected the substance he was testing contained barium sulfate and calcium fluoride, otherwise known as the minerals "barite" ($BaSO_4$) and "fluorite" (CaF_2). This can be seen in his hardness test at the beginning of the article, where he compared the unknown mineral specifically to them, and in his choice of experiments to determine what elements the mineral contained. Suspecting the presence of barite and fluorite was logical, since both can present a grayish-white appearance and they frequently form together in what, today, are called "hydrothermal veins." This term, and the understanding behind it, had yet to be developed in Smithson's time, but he clearly envisioned something like it as having produced the new mineral.

In any case, having successfully identified three of the four elements that he expected to find in the mineral (that is, sulfur, barium, and fluoride), he now turned to the fourth: calcium. After first treating a sample to remove any foreign matter, Smithson subjected it to sulfuric acid (H_2SO_4) to remove the fluoride and let it dry. He then added diluted hydrochloric acid ($HCl + H_2O$) and heated it again. The result, as he described it, was that "on evaporating this solution, a large quantity of sulphate of calcium in crystals was obtained" ($CaSO_4$). Since the calcium could only have come from the sample, this proved it was present in the mineral. Smithson concluded: "From these results, sulphuric acid, fluorine, barytes, and lime, appear to be the elements of this mineral. It is consequently inferable that its proximate principles are sulphate of barium [barite] and fluoride of calcium [fluorite]."

After showing that the new mineral consisted of these two substances, all that was left was for Smithson to measure their proportions, and he described that process in his article. This concluded Smithson's analysis, one of the more detailed of any that he published. The fact that he lavished so much attention on this mineral and had looked for years for another example is striking, and we finally see why in the article's conclusion. He wrote: "This mineral presents us with a remarkable case of [chemical] combination; that of a neutral salt with a body which is not a salt, but belongs to an order which is analogous to metallic oxides," and he spoke of it as "deserving of particular attention from consisting of *four* matters." For Smithson, then, this was a "philosophical" mineral, a rare substance that might offer insights into the nature of chemical combination. What he did not say, but must have been thinking, was that this new mineral (if it *was* a true mineral and not just a combination) provided an answer to the four-element sulphurets, like the sulphuret from Huel Boys, that had been so problematic for the theory of binary chemical combination he had proposed nearly seventeen years earlier.

Had Smithson waited all this time in the hope of using this new mineral as an answer to the critics of his theory? Is this why he lavished so much attention on it and had searched so long in the hope of finding it elsewhere? It would seem that it was, because in the conclusion he mentioned a similar mineral he had investigated. "I have met with another instance of the same kind [of combination]," he wrote. "I have examined transparent crystals which were composed of anhydrous sulphate of calcium and chloride of sodium." Unfortunately, he failed to provide any details, but the fact that such a combination could produce crystals seemed significant. The fluorite/barite mineral did not seem to have that property.

Smithson's article introduced a new and unusual material, but that was all. Despite its apparent promise, Smithson's mineral ultimately failed to present any

From these results, sulphuric acid, fluorine, barytes, and lime, appear to be the elements of this mineral. It is consequently inferable that its proximate principles are sulphate of barium and fluoride of calcium.

The following experiments were made to obtain some idea of the proportions in which these two compound components of this mineral exist in it:

5.6 grs. of this stone in powder were heated in a platinum crucible in so large a quantity of sulphuric acid as to be entirely dissolved. The mixture was then exhaled dry, and ignited. The weight was now 7.85 grs. The increase had, therefore, been as $\frac{40}{100}$.

This augmentation of weight could arise only from the change of the fluoride of calcium into sulphate of calcium.

To know to what quantity of fluoride of calcium it corresponded, two grs. of pure fluoride of calcium in subtile powder were treated with sulphuric acid till the augmentation of weight ceased. The two grains had then become 3.65 grs.; accordingly the augmentation of weight was $= 1.65 = \frac{83}{100}$.

This Derbyshire mineral, therefore, consists of

Sulphate of barium	-	-	51.5
Fluoride of calcium	-	-	48.5
			100.0

Smithson's determination of the relative proportions, by weight, of "sulphate of barium" (barite) and "fluoride of calcium" (fluorite). His determination that the four elements in the mineral first combined to make these two substances fit well with his long-held belief that chemical combination is binary. Courtesy of the Smithsonian Libraries.

new insights. Smithson had clearly hoped to say more about it—which is likely why he waited so long to write this article—but what he wrote was of sufficient interest to be published in the *Annals of Philosophy* and noted in several other journals, and so it entered the scientific literature. The question of whether Smithson's mineral was merely a mixture of the two minerals or perhaps something more intimate, as he suggested, was revived in 1873 when two German researchers reported having grown "individual crystals of fluor spar and barium sulphate," and it was reported that "the authors are of the opinion that the 'Baryte-fluate of Lime' of Smithson may after all prove to be a mineral species and not a mere mixture as many mineralogists assert." Other authors continued to be skeptical. "That of Smithson referred to as Flussbaryt mineral from Derbyshire is probably only a very intimate mixture of fluorite and barite," wrote Friedrich Naumann in *Elemente der Mineralogie* (1874). Ultimately, Smithson's goal was to raise questions and advance scientific understanding, and in this he seems to have been successful. In the early twentieth century the nature of barium and the complexes it can form were still a topic of scientific interest, and Smithson's name continued to be included in that discussion.

Barium was a material of interest throughout Smithson's lifetime. He first encountered it in 1784, when Joseph Black showed him a piece of *Terra Ponderosa Aërata* (barium carbonate). The following year, after going to great lengths to procure a sample for himself, Smithson used it to impress the London scientific community and gain membership in his first scientific society. Over the course of his career he worked with at least two other barium minerals, barium sulfate and barium chloride, using the later in his analysis of "a sulphuret of lead and arsenic." Smithson had great experience working with barium compounds, and in 1823 he put that experience to use in his article "A Means of Discrimination between the Sulphates of Barium and Strontium." He was motivated to write this article because of a report about such a test that appeared in the Royal Institution's *Quarterly Journal*—a test that, in his experience, would not work:

> 19. Test for Baryta and Strontia.—*Baryta and strontian may readily be distinguished from each other by the following process:—Make a solution of the earth, whichever it may be, either by nitric, muriatic, or some other acid, which will form a soluble salt with it; add solution of sulphate of soda in excess, filter and then test the clear fluid by subcarbonate of potash; if any precipitate falls the earth was strontia, if the fluid remains clear it was baryta.*

This was not a trivial matter. The alkaline earths "baryta" (BaO, barium oxide) and "strontia" (SrO, strontium oxide) are similar in both appearance and chemical properties and are frequently found together in mineral deposits, so it was important to be able to tell them apart. In his 1820 article, Smithson had used barium's crystals to identify it. This was also the method recommended in this article, but not everyone was as knowledgeable about crystals as Smithson. Most chemists were more comfortable using "wet" methods in their analysis.

The wet test described in the *Quarterly Journal* recommended first treating the mineral sample with an acid to make it liquid, converting the liquid to a sulfate (with "sulfate of soda"), and placing it in water. Because sulfate of strontia was thought to be soluble in water but sulfate of barium was not, the final step of the test was simply to filter the solution and add "subcarbonate of potash" to the liquid. If sulfate of strontia was present, it would form a precipitate. If nothing precipitated out of the solution, then the sample contained baryta.

The test seemed reasonably straightforward, and, as Smithson noted, "If sulphate of strontium did possess the solubility in water there implied, this quality presented a ready method by which mineralogists would be enabled to distinguish it from sulphate of barium." Unfortunately, in his experience, sulfate of strontium was not soluble in water, and so the test would not work. Instead, Smithson recommended three other tests. The one he preferred made use of crystals. When the sample had been treated to remove other impurities and was "in a state to be soluble in an acid," he wrote, "a more certain, I apprehend, and undoubtedly a much easier proceeding, is to put a particle into a drop of marine acid [HCl] on a plate of glass, and to let the solution crystallize spontaneously." If barium was present it would make "rectangular eight-sided plates," but if strontium was present it would make "fibrous" crystals that were "immediately distinguishable" from those of barium.

The crystal test was simple and unequivocal, but not everyone had his familiarity with crystals. He also had a wet chemical test, and he started to describe it—but then caught himself. "But this process," he wrote halfway through the description, "requires more time and trouble than is always willingly bestowed, and may even present difficulties to a person not familiarized with manipulations on very small quantities." The last thing he wanted was to propose a test that was too difficult, so he moved on.

The third test was something he had thought of just a few months earlier—a simple and straightforward flame test. Some of the sulfate was ground into a powder, combined with "chloride of barium," and then melted into a mass. A piece of this was then placed in alcohol and ignited. If the flame had a bright red color, the mineral contained strontium.

That was the end of the article, and Smithson closed without any further remark. Nevertheless, his article received some notice. Summaries of it appeared in the *Quarterly Journal* (1823), the *Archives des découvertes* (1824), and the *Bulletin des Sciences Mathématiques* (1830). Although the crystal test does not seem to have been adopted, the flame test for strontia began to appear in textbooks. But since the fact that strontium burns with a red flame had long been known, Smithson's name was not associated with it. Instead, what received the most attention was Smithson's remark that sulfate of strontia was not soluble in water.

The *Quarterly Journal* test that Smithson challenged had first been described in 1793 by Thomas Hope, the chemistry lecturer at the University of Glasgow, and it had long been the standard method of distinguishing between baryta and strontia. So it came as a surprise when a fundamental assumption of that test—that sulfate of strontia was soluble in water—was challenged by a chemist of Smithson's reputation and then seemingly supported by another leading chemist, Thomas Thomson. Even more puzzling was that other chemists reported using the test successfully and with excellent results.

The puzzle was solved a few years later, when it was determined that sulfate of strontia is only slightly soluble in water. Today, it is known that it takes 8,800 parts of water to dissolve one part of sulfate of strontia, which seems to explain Smithson's observation. As a longtime advocate of microchemistry, he was used to working with extremely small samples, and this applied to his wet chemistry as well. He does not reveal in his article how much water he used to soak his sample, but it probably was no more than an ounce or two, which may not have dissolved enough sulfate of strontia to be detected. By contrast, chemists who reported that the test worked undoubtedly used more water, and they also reported heating it for many hours, which would have concentrated the solution.

Although never of broad interest, the story of Smithson's challenge became a cautionary tale for anyone working with the alkaline earths, and it led to the development of tests that abandoned the use of water and instead used liquids like ammonium acetate to dissolve strontian compounds. It also earned this article a place in the scientific literature, and, as late as 1916, Smithson's name still appeared in discussions of strontium's solubility.

Taken together, these articles say much about Smithson. They document his skill, both in the field and in the laboratory, his attention to detail, and his intimate knowledge of the physical world. We also see how important his scientific work was to him personally, in willingly spending years and even decades working on a problem in the hope of a new insight. Persistence is a virtue not always associated with Smithson, but he had it in abundance.

———————+———————

14.

A NEW TEST FOR ARSENIC AND "SMITHSON'S PILE"

IN 1822, James Smithson wrote an article describing a method he used for detecting arsenic. Curiously, he had written about this test just a few years earlier, and once he had written on a subject it was unusual for him to revisit it. But in this case, he explained, the "importance of the subject" justified it. The subject was the need to identify the agents in cases of suspected poisoning, and it was a problem that many chemists were struggling with. Poisons were an increasing danger in the early nineteenth century, not only in cases of deliberate poisoning but also from the growing problem of environmental pollution. One of the unintended effects of the Industrial Revolution had been to bring vast quantities of toxic materials into contact with humans, and one of the challenges of Smithson's time was to find ways to deal with this. Two of the most problematic substances were arsenic and mercury, and in his article, Smithson described sensitive new methods to detect each of them.

The bulk of the article was devoted to the test for arsenic. In 1817, the noted physician and toxicologist Mathieu Orfila had written that "the preparations of arsenic are, of all the poisonous substances in the mineral kingdom, the most fatal." And yet in England, where arsenic was widely used as a pesticide, it was essentially unregulated and readily available.

Arsenic, the twentieth most common element in the earth's crust, is gray in its "elemental" form, with a metallic appearance. Surprisingly, in this form it is not poisonous. Arsenic is only poisonous when combined with other elements, and the form usually used as a poison is a combination of arsenic and oxygen commonly known as "white arsenic." Known to the public in Smithson's time as "ratsbane" and to chemists as "arsenious acid," it has no distinctive taste or smell. When ground into a powder it looks very much like sugar or flour, so it could be added to foods and beverages without arousing suspicion. It is not very soluble in liquids, however, so as a poisoned food or drink cooled the arsenic often settled out of solution, giving it a gritty texture, one of the few indicators of its presence.

By the time this happened, the victim had usually consumed some of the arsenic, and it did not take much to be fatal. In Smithson's time substances were generally measured in grains, a grain being the average weight of a single grain of wheat. Depending on a person's physiology, a lethal dose usually consisted of no more than four or five grains, although death could result from as little as two.

The initial symptoms of arsenic poisoning were acute abdominal pain and rapid pulse, followed by convulsions, and then severe vomiting and diarrhea, often leading to death. Arsenic was the chief example of what early toxicologists called an "irritant poison," and it was a particularly unpleasant way to die. Autopsy typically revealed severe damage to the digestive tract, and it was common to find both the stomach and intestines so inflamed that they assumed a deep red color, their inner surfaces covered with mucus and blood. Other organs could also be affected, depending on the individual, the amount of poison consumed, and how it was administered.

With such distinctive symptoms, it seems like the use of arsenic would have been simple to determine, but even when suspected its actual use was difficult to prove. One of the most egregious examples was the case of Robert Donnall, an English physician who fell into debt and who, in the fall of 1816, was charged with murdering his wealthy mother-in-law. English newspapers avidly followed the case and reported that she had fallen ill on two separate occasions, each time after drinking tea he had brought her, and that she died the second time after fourteen hours of vomiting. As the town's medical expert, Donnall urged that she be buried quickly, but before that could happen the authorities received an anonymous letter accusing him of murder. It was reported that when he was shown the letter, "His hands trembled, and . . . it dropt from his hands upon the floor." Under these suspicious circumstances, and despite the doctor's objections, an autopsy was ordered, and the victim's stomach was found to be inflamed in a way consistent with arsenic poisoning. Before it could be tested,

Donnall—who was inexplicably in the room—"accidentally" dropped the organ into a partially filled chamber pot, conveniently compromising the evidence. Nevertheless, chemical tests found arsenic in the victim's last meal, and the doctor was charged with murder.

His trial the next spring was attended by huge crowds. The evidence initially seemed overwhelming, but Donnall mounted a vigorous defense. For example, to explain the inflammation of the victim's stomach, the defense brought in three expert witnesses to testify that it could have been caused by *cholera morbus*, a rare form of dysentery. Although highly unlikely, the prosecution's examining physician had to admit that this was at least a possibility. They also challenged the chemical tests that had detected arsenic. The victim's last meal had been "smothered rabbit"—rabbit stewed in onions. A defense witness described a test in which he had put sliced onions and meat, but no arsenic, in a pot which he let stand for several hours. Using the same two chemical tests as the examiner, he reported that they both incorrectly indicated the presence of arsenic. It was devastating testimony, and when the case went to the jury, the ability of chemical analysis to reliably detect arsenic seemed so suspect that it took them only twenty minutes to find the defendant not guilty.

The problem was that the available tests for arsenic were all indirect. The ideal test would have extracted pure arsenic directly from the evidence, but such a test did not yet exist. The chemical processes used to do this required a large, solid sample and were rendered unreliable by the presence of organic compounds—and, of course, some kind of organic material was almost always present in the evidence of poisoning cases. As a consequence, the liquid tests actually used in these cases sought not to isolate pure arsenic, but rather to produce a distinctive sign of its presence. They worked by bringing the evidence into contact with a chemical solution that produced a precipitate of a specific and characteristic color when arsenic was present.

However, it seemed that every test developed for this purpose had some problem. For example, one of the most common tests used a solution of copper sulfate to test food suspected of being poisoned. If there was arsenic in it, a lively grass-green precipitate known as "Sheel's green" (copper arsenate) would form. However, it turned out that if there was onion juice in the food and if that juice had come into contact with copper, such as from a cooking pot, this too would produce a green precipitate when tested. This is why the "smothered rabbit" in the Donnall case had tested falsely for arsenic. Every test the chemists proposed seemed to have a similar weakness.

Textbooks of the time contain elaborate protocols for dealing with these uncertainties, the first of which was to use multiple tests, each of which had to give a positive indication of the presence of arsenic. They also recommended treating the sample ahead of time to avoid known problems. For instance, if the sample contained tea or coffee, the tannin would affect the results and needed to be removed with gelatins. If the sample contained oil, it needed to be boiled and the oils removed by capillary action from "wick threads." If it contained resins, these needed to be removed with turpentine—but not alcohol, which would dissolve the arsenic. Other substances, such as milk, would change the color of the tests' precipitates or make them difficult to see. In these cases some authors recommended that the sample be treated with chlorine, to "decolorize" it, but this could create new problems by affecting the colors of the tests' precipitates.

In 1821, with the English public increasingly fearful of arsenic poisoning and the reliability of the tests for it widely questioned, the Society for the Encouragement of Arts, Manufactures and Commerce offered a gold medal to any person "who shall discover to the Society a test for arsenic in solution, superior to any hitherto known." The announcement brought this issue to the attention of the scientific community, and although he did not apply for the prize, the following year Smithson published this article describing how to apply his arsenic test to poisoning cases. Smithson emphasized his test's adaptability, extreme sensitivity, and the fact that it could detect arsenic in many different forms: as an acid, in combination with oxygen, and in its metal-like elemental form.

The test's first step was to combine the sample with "nitrate of potash" (potassium nitrate, KNO_3) and heat it with a flame until it dried and caught fire. This burned away the organic material and caused any arsenic to combine with the potassium. Now any excess potassium needed to be removed, and this was done by carefully adding vinegar until the solution was neutralized. This liquid was then dried and dissolved in water, which put the "arsenate of potash" (K_3AsO_4) into solution. Then, just as he had in his 1819 article, Smithson described how adding a small amount of "nitrate of silver" (silver nitrate, $AgNO_3$) to the liquid would cause a "brick-red precipitate" to appear—but only if the original sample had contained arsenic. The distinctively colored precipitate that settled in the bottom of the liquid was "arsenate of silver" (Ag_3AsO_4), and even a tiny amount of arsenic would make a lot of it. He reported that a grain of arsenic acid would make 4.29 grains of the precipitate, that a grain of arsenic oxide would make 4.97 grains of it, and that a grain of pure arsenic would make 6.56 grains. This, along with the color, made the appearance of the precipitate hard to miss, and it was an important part of the test's sensitivity.

Smithson's test was based on the reaction of silver nitrate with compounds containing arsenic, but it wasn't the first arsenic test to make use of this chemical. One of the tests used in the Donnell case had also used silver nitrate, but in that test any arsenic in the sample would combine with it to produce "*arsenite* of silver" (Ag_3AsO_3), which is a yellow precipitate. What they discovered in the Donnell case was that even when no arsenic is present, the naturally occurring alkaline phosphates found in animal tissues will also combine with silver nitrate to produce a yellow precipitate—and thus give a false positive for the poison.

The fact that Smithson's test produced a distinctive brick-red precipitate, together with his method of removing any organic material in the sample, was widely credited as having addressed the problem of using silver nitrate. His test was quickly adopted as one of the standard methods of detecting arsenic, and a positive result was a strong indication that the analysis was correct. Finding a reliable method of extracting elemental arsenic from the sample would continue to be a goal, but until such a method was found Smithson's test served the purpose. As the distinguished chemist Richard Phillips noted in a widely cited article, "from repeated trials, I consider the confirmatory evidence afforded by this experiment as amounting almost to demonstration." There were other expressions of support as well. At the end of a long discussion of arsenic in one of his textbooks, Robert Hare concluded, "Mr. Smithson's plan . . . seems to me an excellent one." In his widely reprinted *Elements of Medical Jurisprudence* (1825), Theodric Beck noted with approval that Smithson's plan was "well calculated to detect the most minute portions of poison." Smithson's article was also reprinted in the United States where, as in England, arsenic poisoning was a significant concern.

Smithson's test was quickly adopted, and for more than fifteen years it was one of the most reliable and widely used methods for detecting arsenic. But in 1836 the chemist James Marsh announced a new test that was finally able to meet the criteria of the Society for the Encouragement of Arts competition and earn its coveted gold medal. The "Marsh Test," as it came to be known, was based on a new principle and used gaseous hydrogen to extract pure arsenic from a solution and deposit it on a glass plate. This was what everyone had been waiting for, and the test almost immediately became the standard method of arsenic detection. It remained in widespread use until the 1970s, when it was finally replaced by the new technologies of chromatography and spectrophotometry. As for Smithson's test, like most of the other "liquid" arsenic tests, its use declined rapidly once the Marsh Test became available, and by midcentury it had been largely forgotten.

But Smithson's article also described another test, this one for mercury, that would remain in widespread use for nearly a century. Mercury, like arsenic, is a poisonous metal whose toxicity varies according to its chemical form. Pure mercury is a liquid at room temperature, and in that form, it is so heavy that it is not readily absorbed—although it is still a health hazard. However, mercury acquires different properties when combined with other materials. It is toxic in the form of "corrosive sublimate" ($HgCl_2$), and somewhat less toxic in the form of "calomel" ($HgCl$). Indeed, in Smithson's time pure mercury and calomel were better known as medicines than as health hazards. At that time, mercury was one of the few treatments that seemed to provide relief from syphilis. "Blue mass" pills, which contained either calomel or liquid mercury mixed with chalk along with taste and coloring agents, were used by physicians to treat it and a wide variety of other medical conditions.

This is not to say that doctors in Smithson's time were unaware of the hazards of mercury exposure. In 1811, Andrew Mathias, a member of London's Royal College of Surgeons and surgeon extraordinary to Queen Charlotte, wrote *An Inquiry into the History and Nature of the Disease Produced in the Human Constitution by the Use of Mercury*, in which he accurately described the effects of mercury poisoning, calling it "the disease of the remedy." Like his contemporaries, however, he saw these symptoms mostly as the undesirable side effects of an effective medicine. His book was not intended to warn about the dangers of using mercury, but rather to guide physicians in the best ways of using it while limiting its undesirable effects.

The use of mercury in manufacturing was also known to be hazardous. Gilders, for example, had long added mercury to gold to make it soft enough to apply to metal objects, which were then heated to vaporize the mercury and leave a decorative layer of pure gold. These workers knew well the hazards of mercury vapors, one of the first symptoms being a continuous, uncontrollable salivation. As a result, gilders made every effort to control these fumes as they worked, but the obvious solution of directing them up chimneys caused other problems. In 1810 it was reported that Birmingham chimney sweeps were reluctant to clean the chimneys of gilders because exposure to the dust "resulted in long salivations." These hazards affected a relatively small number of people, though, and so were generally classified (and dismissed) as examples of the "diseases of artisans." Nor was there any attempt to keep workshops using mercury away from residential areas—at least those areas occupied by residents with low incomes.

But, beginning around the end of the eighteenth century, Europeans found pollution increasingly difficult to ignore. One of the effects of the Industrial

Of Mercury.

All the oxides and saline compounds of mercury laid in a drop of marine acid on gold with a bit of tin, quickly amalgamate the gold.

A particle of corrosive sublimate, or a drop of a solution of it, may be thus tried. The addition of marine acid is not required in this case.

Quantities of mercury may be rendered evident in this way which could not be so by any other means.

This method will exhibit the mercury in cinnabar. It must be previously boiled with sulphuric acid in the platina spoon to convert it into sulphate.

Cinnabar heated in solution of potash on gold amalgamates it.

A most minute quantity of metallic mercury may be discovered in a powder by placing it in nitric acid on gold, drying, and adding muriatic acid and tin.

A trial I made to discover mercury in common salt by the present method was not successful, owing, perhaps, to the smallness of the quantity, which I employed.

I am, sir, yours, &c.,

JAMES SMITHSON.

Smithson's description of his new test for mercury. Consisting of only nine sentences, it appeared almost as an afterthought at the end of his 1822 article. Although he did not mention it, the test utilized the newly discovered "galvanic force." Courtesy of the Smithsonian Libraries.

Revolution was the concentration of labor and similar types of businesses. Even when not accompanied by large scale mechanization, this concentration offered significant efficiencies in transportation, supply of materials, and access to skilled workers, and by the beginning of the nineteenth century similar types of manufacture were being concentrated in specific geographical areas, often in cities.

Unfortunately for the residents of Paris, the majority of the businesses drawn to their city were producers of luxury goods. Paris had a long history as a center of luxury and fashion and was known for having a well-established community of skilled workers and specialized businesses able to produce elaborately worked goods from exotic materials. A great number of these trades used some form of mercury in their work. Gilders were among these, but gold foundries also used it, as did mirror makers and tool makers, particularly those making scientific instruments. Mercury in its mineral form (HgS) was used to color the wax seals on government documents and make dyes. Another form, "corrosive sublimate" ($HgCl_2$), was added to ink to keep it from getting moldy. This toxic powder was also widely used as an insecticide; adding it to the paste used to bind books kept insects away, and sprinkling it on rugs and furniture killed insects without harming the fabric. Mercury was also used in large quantities by Parisian chemical firms, who made pharmaceuticals, dyes, and other chemicals with it. And, of course, Parisian hospitals used mercury on their patients. But of all the trades, hatmakers used the most mercury, and they used a particularly dangerous form—mercury nitrate ($Hg(NO_3)_2$).

The majority of these workshops were located on the right bank of the Seine, primarily in and around Paris's historic Le Marais district. But this was not the picturesque Marais seen today, which grew out of the city's renovation in the 1850s and 1860s. In the 1820s, when Smithson knew it, the area was little changed from medieval times, with impossibly narrow streets, overcrowding, poor sanitation, crime, and squalid poverty. In spite of that, it was also an increasingly busy commercial area, with hundreds of new shops and businesses opening every year.

The concentration of workshops into one area of the city meant that much of the pollution they produced was concentrated there as well. As the number of businesses grew, so did the amount of mercury used and pollution produced. Before the French Revolution, only a few hundred pounds of mercury had been brought into the city each year. By 1810 Paris's annual consumption of mercury had risen to ten tons, and by the early 1820s, when Smithson wrote his article, it had increased to nearly twenty tons a year. Although it was used in different forms and for different purposes, most of it ended up being disposed on the right bank. Whether dumped directly into the river or into a canal that drained into the river, allowed to seep into the soil, or discharged into the atmosphere in the form of toxic fumes and dust—which then settled on the houses of nearby residents—one way or another, most of the mercury brought into Paris during this period ended up in the city's air, ground, and water.

The practices of the hatmakers are a good example of how this happened. Hats were one of Paris's most important products in the early nineteenth century, as items of fashion or in huge numbers for Napoleon's armies, and the hats preferred by everyone were made from molded felt. Felt, in turn, was made from the hair of beavers, rabbits, or hares, which was first scraped from the skins, then compressed and skillfully shaped into a hat. Soaking the animal skins in a solution of mercury nitrate made the hairs easier to remove. It also changed the hairs themselves, making them much easier to work with and producing a denser, more uniform felt.

This form of mercury was particularly toxic, and once the furs had been treated with it, they discharged mercury through the rest of the manufacturing process. It started with the furs being soaked in a solution of mercury nitrate and then dried, exposing the workers not only to the solution, but to its fumes. Any solution that spilled or became too dirty would have simply been poured on the ground—or in the street—where it would contaminate the soil and groundwater. Once the skins were dry, the hair was scraped off and collected. The skins, still contaminated with mercury, were sold for other purposes, while the hairs were formed into felt. The hairs were very light, and to prevent them from blowing away, they were worked in closed, breezeless rooms. Each hat started off as a large oval pile of hair, about four feet long, which was carefully cleaned and slowly worked into shape by a skilled journeyman and at least one assistant. It was dirty, dusty work, and visitors to the shops reported not being able to see anything that was more than eight feet away.

As the felt was compressed, it was repeatedly treated with hot water, placed in molds, and heated. Each step removed a bit more of the mercury nitrate, and it appears that the finished hats did not pose a health risk. This means that all the mercury used in making them was discharged into the environment. It has been reliably estimated that each journeyman hatter used around ten kilos of mercury nitrate per year, and by the 1820s Paris had the largest concentration of hatters in the world: nearly 150 hat-making firms, employing between two and three thousand workers, and producing nearly two million hats each year. A simple calculation shows that as much as sixty thousand pounds of toxic mercury nitrate was being dispersed into the city each year—year after year—by just one trade.

Eventually the problem became too severe to ignore. In 1818, just four years before Smithson wrote his article, at least three thousand Parisian workers were reported to be sick from mercury poisoning. Whole families of the city's residents exhibited symptoms of poisoning. In 1821 a Health and Safety Council report called attention to "the skin diseases that are so common in Paris, especially in the indigent classes," and blamed it on their washing clothes in mercury-polluted water. In the 1820s, government-encouraged changes to the workshops began to improve conditions for the workers, but because most of these changes were to enhance ventilation, the improvement was at the expense of nearby residents, whose exposure to mercury was now dramatically increased.

The pollution problem continued to grow. Between 1770 and 1830 it is estimated that a total of six hundred tons of mercury escaped into the Paris environment, most of it on the right bank. The scale of the problem and the many different forms of mercury flooding the environment overwhelmed the city's inchoate public health system and finally, beginning around 1827, things came to a head. Paris was struck by an epidemic that would eventually affect as many as forty thousand residents. The disease particularly afflicted the poor, and in the early stages, victims suffered from painful blisters and sores on their hands and feet. As the disease progressed there could also be vomiting, colic, and diarrhea. Most victims eventually recovered, but the experience left them weakened and susceptible to repeated sickness when they resumed their lives. Today the "epidemic" is widely thought to have been an outbreak of acrodynia, a condition caused by exposure to heavy metals, especially mercury.

Paris was eventually able to control the use of mercury and to decontaminate the city, but that important work took place after Smithson's death. He did not mention the mercury problem in his article but, living in Paris in the early 1820s, it seems impossible that he could have failed to be aware of it. He was simply too observant and inquisitive to have missed the smoke and foul smells that poured into the streets from the shops in the right bank or the sickly cloud that always seemed to hang over that part of the city.

Smithson's awareness of the problem explains his decision to include the mercury test in his 1822 article, which was otherwise about detecting arsenic. Although he described it only briefly, his test became the most practical and sensitive method for detecting mercury yet developed. Unlike the traditional chemical tests that it replaced, Smithson's test did not depend on a distinctive chemical reaction or color change. Instead, it employed an entirely new principle: the recently discovered galvanic force.

Electricity was a well-established branch of science when Smithson enrolled at Oxford, and he undoubtedly attended lectures on it in his natural philosophy course. When people talked about electricity in the 1780s, they meant what we now call "static electricity." Smithson would have learned that electricity consists of positive and negative particles, and that "like" particles repel each other while "opposite" particles attract. He would have also seen demonstrations at Oxford in which "electric machines" were used to collect and store these particles. Once a sufficiently large charge had been collected, it could be used to power ingenious devices or discharged as a single, impressive spark. Although it had relatively few practical applications, this form of electricity was philosophically extremely interesting. Indeed, the fact that heating certain kinds of crystals causes them to develop electric charges suggested a fundamental connection between heat and electricity. Smithson himself was interested in this unexpected connection and made an extensive study of "electric" crystals early in his career—a study that he shared with colleagues, but unfortunately never published.

Most of what Smithson learned at Oxford belongs to a branch of electrical research now called electrostatics. In the late 1780s, near the time that Smithson graduated, "current" electricity was first observed. That is when the Italian physician Luigi Galvani began a series of experiments after accidentally causing the leg of a dissected frog to kick. Galvani found that the kick was caused by electricity produced by simultaneously touching the leg with two different metals that were also in contact with each other. It was a remarkable discovery, and this form of electricity is now called a "galvanic" current in his honor. But it was the Italian physicist Alessandro Volta who correctly identified it as being based on a chemical reaction, and Volta used that understanding to build the world's first "battery."

In 1800 Volta published a detailed description of this new chemical device. His experiments had shown that the best metals to use were silver and zinc, and that while water would work as the conductor (in place of the frog leg), salty water worked better. Accordingly, Volta's battery was built from silver coins, zinc discs, and round pieces of flannel soaked in brine. The basic unit was one of each: a single silver coin with a wet flannel disc on top, and a zinc disc on top of that. Putting these three things in contact produced electricity, but the amount of electricity that each of these "cells" produced was quite small. Fortunately, Volta also discovered that if several cells were stacked (or "piled") on top of each other, their power would be combined.

As long as the order of the components stayed the same, there seemed to be no limit to the number of cells that could be combined in this way—or to the power they could produce—and Humphry Davy was soon using larger, improved batteries at the Royal Institution to dissociate materials that were previously thought inseparable. Volta attributed this power to separate materials to what he called the "electromotive force." He imagined this force as a kind of invisible fluid that carried electricity along with it as it flowed through conducting materials, and he used the same analogy to explain another remarkable property of galvanic electricity—its ability to move matter.

One of the things the early researchers discovered was that deposits would soon build up on certain parts of batteries, reducing their power and making them unreliable. This problem got worse when they started using acidic solutions instead of water or brine. Acidic conducting solutions helped activate the battery

A "Voltaic pile." Volta's invention of what is now called the "battery" made a new kind of electricity available to the scientific community. Interestingly, the battery's design was strongly influenced by his studies of the anatomy of the torpedo fish, which has a specialized organ capable of delivering an electric discharge. Indeed, the name he proposed for this new device was "artificial electric organ." The common name for any form of battery, however, quickly became a "pile." Courtesy of the Smithsonian Libraries.

and increase its power, but they also increased the deposits accumulating on the battery's ends. These deposits were mostly metals, impurities that had dissolved in the acid, and as soon as the battery's current started, these materials would migrate toward one of its terminals. The fact that acids and alkalis were attracted to different terminals seemed to suggest that the distinction between them was at least partly electrical. This in turn implied some connection between electricity and the chemical bond, and it was such an interesting idea that Smithson's friend Berzelius soon developed it into an entire chemical theory.

Batteries and electrical attraction became important research topics in the first decades of the nineteenth century, and much of what was learned about these topics came from a type of experiment called the "galvanic circle." Unlike the dramatic, and expensive, electrical experiments that Humphry Davy conducted, these were minimalist experiments intended to reduce the battery to its essential components and then study the effect of making very specific changes to the materials being used. Because all that is required to make a battery are two different metals and a conducting liquid, these experiments could be made on a very small scale, with just a few drops of acid and two different wires—which was fortunate, since there were an almost unlimited number of materials to test. Among other things, galvanic circles were used to study how metals accumulated on a battery's terminals. Acids in which various substances had been dissolved were tested with "piles" of various construction to see which substances would be deposited on which metals. This line of research eventually led to the development of electroplating.

Although he did not identify it as such, the mercury test that Smithson described in his article was clearly a galvanic circle. He wrote, "Compounds of mercury laid in a drop of marine acid on gold with a bit of tin, quickly amalgamate the gold"—which fulfills all the necessary conditions of a simple battery. The two metals were gold and tin, and the conducting liquid was an acid, with a mercury compound dissolved in it. In practice, the test is almost magically simple. If a few drops of hydrochloric acid (with mercury dissolved in it) are placed on piece of glass, and the ends of two wires—one gold and one tin—are placed in the acid, a normal chemical reaction is produced. The acid will begin to dissolve the tin, meaning that small bubbles will begin to appear around the part of the tin wire touching the acid, while the gold will be unaffected. But if the ends of the two wires not touching the acid are brought together, this reaction stops—because a battery has been made. Bringing the two wires together causes an electric current to flow through the acid, from the tin to the gold, which can be seen by bubbles now forming around the gold wire, not the tin one. After a short time, when the gold wire is removed from the acid, its end will be covered with a layer of mercury. This is the essence of Smithson's test.

Smithson's ingenious use of the galvanic circle was one of the earliest practical applications of the new galvanic force. It is noteworthy that Smithson was the first to propose this test, because his name has not previously been associated with electrical research. The nine sentences devoted to this test constitute the sum of all of his published works on electricity, but they show that he was familiar with this line of inquiry and actively incorporating it into his scientific work.

One of the advantages that Smithson enjoyed by living in Paris was the opportunity to interact with many of the premier scientific figures of his time. In his article on "Capillary Tin" (1821), he mentions having recently had a conversation with André-Marie Ampère. Ampère would soon become famous for discovering "Ampère's law," a finding so important that the *ampere*—the fundamental unit of electric current—is named after him. Just two years later in 1823, Smithson met with Hans Christian Oersted, who first demonstrated the connection between electricity and magnetism. Oersted was visiting Paris and later wrote to his wife that they spent three hours together, and Smithson had demonstrated his remarkable microchemistry. Smithson's interest in electricity appears much more profound than has previously been thought. At age fifty-seven, when this article was published, he was clearly keeping up with the latest discoveries and actively engaging with some of the leading electrical researchers.

Smithson's humbly introduced test soon attracted considerable attention. In Germany two reports on it appeared in the *Polytechnisches Journal*, and in America, William Henry quickly included it his popular chemistry text. But French scientific journals showed the most interest. Within a year, reports on it had appeared in the *Annales de Chimie*, *Archives des Découvertes*, *Medical Jurisprudence*, and *Annales des Mines*, which, tellingly, ignored any mention of arsenic and reported only the mercury test.

Smithson's test came at just the right time, but it seems that the only place where he did not receive credit for it was back in England. It was instead given to Charles Sylvester, an English chemist, science lecturer, and inventor. Sylvester had a particular interest in galvanism, and in 1812, fully ten years before Smithson's article, he had described a method of using the galvanic force to detect a variety of metallic poisons—including mercury. Just like Smithson's test, Sylvester's was clearly a galvanic circle, and, like Smithson's, one of the metals it used was gold. The second metal was not tin, though, but instead either zinc or iron, and the only form of mercury Sylvester reported having tested for was "corrosive sublimate" ($HgCl_2$).

Sylvester's article received some attention when it first came out, and it was reprinted in the American journal *The Eclectic Repertory* (1814), but beyond that it failed to attract much interest. Sylvester succeeded in demonstrating that the galvanic force could be used to detect mercury, but he failed to develop that into a simple, practical test, and there is little evidence that what he proposed was actually used. Even by the loose standards of the time, this was a weak priority claim, and Sylvester's test appears to have been largely forgotten until Smithson's article was published. But beginning in the 1820s, as English textbook writers included discussions of the important new test, they chose to credit it to Sylvester. "Mr. Sylvester has proposed another ingenious application of galvanic electricity," declared *The Elements of Medical Jurisprudence* (1825), while *A Practical Treatise on the Use and*

Application of Chemical Tests (1829) praised him for recommending "a galvanic circle, formed of zinc and gold, as an agent for detecting corrosive sublimate." In 1830, the English chemist Edmund Davy submitted an article to the Royal Society about detecting metallic poisons. Davy, mute about Smithson's contribution, praised "the ingenious mode proposed some time since by Mr. Sylvester." When the society published his article in its journal, it seemed to make official that Sylvester should be given credit for the test.

Even if Sylvester was the discoverer, that hardly justified excluding Smithson from any connection with the test. Did this happen because of Smithson's estrangement from the Royal Society? All we really know is that until midcentury, and almost without exception, English textbooks failed to note any connection between Smithson and the galvanic test. This contrasts starkly with the situation elsewhere, particularly in France.

Smithson's reputation on the continent was enhanced in 1829, when the noted toxicologist Mathieu Orfila wrote an article in which he evaluated the mercury test. Although he suggested some precautions when using it, he ultimately endorsed the test, writing that it was so sensitive that "the small apparatus imagined by Mr. James Smithson . . . is able to serve to detect a few atoms of a mercury preparation in a suspect liquor." Orfila's article was quickly reprinted in the *Annales de Chimie*, and his endorsement of the test was widely cited. Abstracts of it even appeared in some English-language journals, including *The Philosophical Magazine*, the Royal Institution's *Quarterly Journal*, and *The North American Medical and Surgical Journal*.

In less than a decade, Smithson's test became the standard method of detecting mercury. It made such distinctive use of materials that, on the continent, his name was almost always associated with it. By the 1840s, the standard name for the test had become "le pile de Smithson"—Smithson's battery. Smithson's reputation received another boost in 1845, when a paper read to the French Academy praised the test as "the only progress of our time" in the detection of mercury. The academy's apparent endorsement of Smithson's importance was widely reported in scientific journals, including several English journals, and from that point forward English references to Smithson increased, while references to Sylvester ceased to appear.

French scientific and medical journals would continue to refer to "L'appareil de Smithson" and "le pile de Smithson" throughout the nineteenth century, while in Germany it was known as "Smithson'schen Vorrichtung"—Smithson's device. It had also become the standard method of testing for mercury in England and North America, and directions for using the pile regularly appeared in textbooks.

Alfred Joseph Naquet's widely used *Legal Chemistry* (1876) devoted three pages to it and informed students that "Smithson's pile" was "the most delicate in use for the detection of mercury." From 1865 until at least 1907, every issue of *The United States Dispensatory* recommended the use of "Smithson's process."

Smithson's method remained in widespread use well into the twentieth century and was almost invariably associated with his name. Given its importance and the attention it received, one would expect the mercury test to have become an important part of Smithson's legacy, but surprisingly it did not. It has hardly been mentioned, even within the Smithsonian Institution. *The Scientific Writings of James Smithson* (1879), the institution's primary assessment of Smithson's work, notes that the mercury test was still being used in the 1840s, but then incorrectly implies that his arsenic test had been more important. There was no mention of "Smithson's pile" (or any of the test's other names) in either *The Scientific Writings* or any of Smithson's subsequent biographies. This is particularly surprising given that *Scientific Writings* was produced as part of the Smithsonian Institution's commemoration of its founder, and it was clearly hoped that some great discovery or invention of his would be discovered.

In thinking about how this omission could have happened, it is important to remember that the *Writings* was assembled on both a deadline and a budget, so the research that went into it was limited—confined to the English-speaking world. Questionnaires were circulated only in the United States and England, and only English-language journals were consulted. This may explain how the test was missed, for the English did not originally credit Smithson with discovering it. Only after 1845 (following the lead of the French Academy) did English-language publications associate his name with the test. But Smithson had died in 1829, and he had not lived in England since 1817. This, combined with his estrangement from the Royal Society, could have meant that by the early 1840s Smithson was no longer a figure of interest to English science. In 1845, when the French Academy credited Smithson for the mercury test, it may not have been obvious to the English that it was James Smithson, the English chemist, that they were referring to. There would have been little reason to think that the English James Smithson had developed what was often considered a medical test or to associate him with anything related to electricity. So, when the French Academy credited Smithson, it may have looked to the English like a French priority claim. The fact that the claim appeared mostly in French journals would have seemed to confirm this, and it did not help that in France he was sometimes referred to as "le docteur Smithson."

This idea of a mistaken identity is an assumption, but it seems to be the best way to account for the changing fortunes of Smithson's reputation. It also explains why the Smithsonian Institution, in its search to celebrate its founder, failed to recognize one of his major achievements. They simply assumed that "Smithson's pile" referred to a different Smithson. So another part of Smithson's story slipped out of history.

15.

SMITHSON'S LAMP
AND THE "SAPPARE"

IN 1822 Smithson began to write about the scientific tools he designed and used. Most of his remarks were asides in articles about other topics, but two of his articles were more focused. One, titled "Some Improvements of Lamps," consisted mostly of observations about oil lamps, and as he began Smithson clearly felt the need to explain why he was writing on such a commonplace topic.

"It is, I think, to be regretted," he wrote, "that those who cultivate science frequently withhold improvements in their apparatus and processes, from which they themselves derive advantage, not deeming them of sufficient magnitude for publication." But, he argued, all useful information should be imparted, "however small may appear the merit which attaches to it." He then made some observations about the wicks used in oil lamps. A wick is a length of braided cotton, much like a string or a rope, that works by capillary action to draw fuel up from a lamp's reservoir to feed its flame. He gave two reasons why the common practice of putting long wicks in lamps was "extremely inconvenient." First, the long wick took up space in the lamp, thereby reducing the amount of fuel it could hold, and second, it tended to collect dirt and other impurities found in the lamp's oil, which could make the flame burn erratically.

Under ideal conditions the fuel in a lamp burns, not the wick, and the oils used today are so highly refined that wicks rarely need attention. In Smithson's time the best available fuels were whale oil, colza (rapeseed) oil, or olive oil, none of which were particularly refined. By modern standards they did not burn cleanly, and this made the flame sputter, which in turn burned the wick, causing the lamp to smoke. When this happened the lamp needed to be extinguished, the wick pulled out a bit, and the burnt parts on the end of the wick cut off. In Smithson's time this kind of maintenance was an almost daily occurrence, and the constant trimming they required was the reason that lamps were furnished with such long wicks.

But Smithson had an idea. He recommended replacing the long wicks with a short metal tube, filled with either a tightly fitting cotton wick or wadded up cotton wool. As long as it could reach the bottom of the reservoir, this cotton-filled tube would draw fuel to the flame just like the wick it replaced. But what happened if the cotton in the tube got burned? Smithson solved this by leaving a space at the top of the tube, which he filled with a loose piece of wick, explaining that "this loose end receives [a] supply of oil from the cotton under it with which it is put into contact, and when it becomes burned, it is easily removed [and replaced]." It was a clever—if somewhat fussy—solution, and although never widely adopted, it appears to have worked.

Now Smithson turned to the problem of fuel. "Oil," he wrote, "is a disagreeable combustible for small experimental purposes, and more especially when lamps are to be carried in travelling." At a time when all land travel of any distance was either on horseback or in horse-drawn carriages, oil was almost certain to spill and leave a greasy film on everything it encountered. For a man like Smithson, who spent large parts of his life on the road, this must have been a constant problem. One possible solution would have been to forego lamps entirely and use candles. Candles provide a workable flame, but that flame constantly lowers as the candle burns down, which is extremely inconvenient for blowpipe work. The blowpipe needs a flame that stays in a fixed position. It was a problem that Smithson solved by designing a lamp that burned wax.

Smithson described two kinds of wax lamps, the first being basically what we now call a container candle. These do not seem to have been common in the early nineteenth century, so Smithson described how to make one. The first step was to prepare the wick, which he did by drawing it slowly through melted wax. After it dried, he placed one end "in a burner made of a bit of tinned iron sheet." This "burner" was then placed in the bottom of a small china cup, and the wick was held vertical while the cup was packed with "fragments of wax" pressed down to hold everything in place.

Fig. 1.

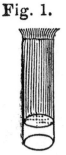

Fig. 2.

Smithson's drawings of the parts of the metal tube assembly he designed to replace the long wick. On the left is the metal tube, shown partially filled with cotton wick, and with a space at the end. (Note that in this drawing the tube is shown upside down, just as it was in Smithson's article.) The image on the right is identified as the loose piece of wick placed in the end of the tube, which, when necessary, could easily be replaced. Courtesy of the Smithsonian Libraries.

This lamp was surprisingly small. The wick was simply a cotton thread, and the whole assembly, including the wax, fit in a china cup "about 1.65 inches in diameter and 0.6 in. deep." Given the tiny size of the samples Smithson worked with, this small flame was all he needed, and a lamp like this offered advantages for a traveler. The lamp's construction meant it could be reused almost indefinitely. As it burned down, new pieces of wax could be added to keep it going, and Smithson tells us that if the wick got burned it could simply be dug out "with a large pin down to the burner, and a fresh bit of waxed cotton introduced."

Smithson informs us that he used a second kind of wax lamp with the blowpipe. While he provided an uncharacteristic amount of detail about the other topics in this article, he chose to reveal almost nothing about his blowpipe lamp. All he wrote was that it "has, of course, a much larger wick, and this wick has a detached end to it."

Despite his failure to describe it, Smithson was careful to provide instructions on his special method of lighting the blowpipe lamp. He wrote that "great care must be taken that each time the lamp is lighted, bits of wax are heaped up in contact with the wick, so that the flame shall immediately obtain a supply of melted wax." What Smithson did not make clear was that once his blowpipe lamp was warmed up, it worked by using some of the flame's heat to melt additional wax to keep it going, and it could keep going until all the wax was consumed. The problem was to get it going in the first place without burning the wick. Packing wax around the wick before lighting it was Smithson's way of preventing this, and he wrote that this was "the great secret on which the burning of wax lamps depends."

Smithson's final remarks addressed the best way of extinguishing a wax lamp, and for this he recommended placing a solid piece of wax in contact with the wick and then quickly blowing it out. "This preserves the wick entire for future lighting again." He also recommended this method for extinguishing ordinary candles, arguing that it was "much preferable to the use of an extinguisher."

Fig. 3.

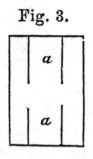

Fig. 4.

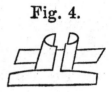

Smithson's drawings of how to make a "burner" for his wax lamps. The image on the left shows the four cuts made on a piece of "tinned iron sheet," and on the right is the "burner" produced when the middle sections were bent up and inward. The wick was held between the bent pieces. Smithson's larger lamp also used this type of burner with a tube assembly. The size of the burner was not specified. Courtesy of the Smithsonian Libraries.

Smithson had clearly devoted considerable thought to the subject of lamps, but the changes he proposed seemed to offer few practical advantages. As a result, reaction to the article was tepid at best. It was briefly noted in a few European scientific journals, and one writer recommended it to travelers—to heat water for coffee and shaving—but there is no evidence that it was actually used.

Smithson failed to describe the wax lamp that may have been of greatest scientific interest—his blowpipe lamp. Other than instructions about how to light it, the only details he gave were that it had "a much larger wick" than his small lamp, and that it had "a detached end to it." Fortunately, this is enough to make some educated guesses about how the lamp could have worked. If the lamp's wick had a detached end, the only way this makes sense is if he was also using a cotton-filled metal tube, similar to what he recommended for oil lamps. And this, in turn, would have needed to be held in place by a metal "burner," similar to the one he used in his small lamp. Studying Smithson's illustrations allows us to visualize how this could have worked.

The flame needs to be elevated in any lamp used with a blowpipe. There has to be enough room for the blowpipe to be positioned on one side, toward the bottom of the flame, and for the sample (along with whatever is being used to hold it) to be on the other side, a bit lower. In Smithson's lamp this requirement was easy to meet: the metal tube holding the wick simply needed to stand above the lamp's container.

As to the container itself Smithson gives no clues, but he mentions that he assembled his other wax lamp in a tiny china cup. Assuming he also used tableware for the blowpipe lamp, one possible container would have been an egg cup.

The author's re-creation of the wax lamp Smithson used with the blowpipe. The wick, brass tube, and metal burner are all embedded in wax in the cup. Once lit and warmed up, part of the flame's heat is conducted down to the burner, where it melts enough wax to keep the lamp going. This lamp has a quarter-inch wick, and, using beeswax as a fuel, it can burn for several hours. The heat it produces and the size of the flame both compare favorably with an oil lamp of similar size. When the flame is extinguished, the wax immediately begins to harden. In a short time, the lamp can be transported with no possibility of a spill. Photograph by James Gleason.

Tests showed that a quarter-inch diameter wick is about as large as is practical with a wax lamp, and that size fits comfortably in an egg cup. Tests also showed that beeswax works the best with Smithson's design. Paraffin, the ubiquitous modern wax, was unknown in Smithson's time, but several plant-based waxes would have been available, notably rapeseed wax. Rapeseed oil was known to burn with a particularly bright (and presumably hot) flame, which made it a likely candidate. In actual use beeswax melts more readily (which makes lighting the lamp easier) and is significantly softer, which makes it easier to work with.

Ultimately, the importance of Smithson's article on lamps seems to lie mostly in what it reveals about him. Instruments were critical to the practice of chemistry, and chemists in Smithson's time were often highly individualistic in their instruments and in the ways they used them. They were reflections of lifestyle and temperament as much as science. Smithson's lamps were optimized for both travel and microchemistry, a form of chemical analysis that he helped establish.

Smithson described another tool he had developed in his article "A Method of Fixing Particles on the Sappare" (1823). This was about ways he had developed to hold mineral samples in the flame of the blowpipe, but there was more to the article than just practical tips. The new methods Smithson described extended the range of what the blowpipe could do and provided a number of new tests that could be used to identify and analyze minerals.

One of the problems that mineralogists faced during Smithson's time was the destructive nature of their tests. Analyzing a mineral could easily use up the entire stock of it, and this was particularly true with chemical tests, which subjected the sample to liquid acids and alkalis. As Smithson wrote, "The chemical method justly boasts its certainty; but it carries destruction with it, and often bestows the knowledge of an object only at the expense of its existence."

In the late eighteenth century, the geologist and sometime mineralogist Horace de Saussure had addressed this problem, proposing a system of identifying minerals by the way they melted. Saussure had developed a system for categorizing minerals by the appearances they presented in the blowpipe flame, and the beauty of his method was that it used very small samples. The problem, however, was finding a way to hold the little bits of mineral in the flame without it simply blowing them away—as it did when the sample was placed on a piece of charcoal. Saussure's initial solution was to embed the sample on the end of a small glass rod, heating the end just enough to soften it, and then gently placing the sample on top. Once the glass cooled, the sample would effectively be glued in place and the rod would serve as a holder, allowing the sample to be securely held in the intense heat of the flame. It was an effective technique, and Smithson reported using it for some of the tests in his analysis of tabasheer. To make the delicate task of mounting the sample easier, Saussure designed a special blowpipe holder that Smithson may also have used.

Saussure would go on to develop other blowpipe improvements, but the most promising was a method he developed for holding samples in the fire. The glass rod worked fairly well, but Saussure found that for minerals with a high melting point, or for minerals that simply needed to be held in the heat for a long time, the glass would soften and the sample sink down into it, making it impossible to continue the test. For these minerals he developed another holding method, one that used a bluish, highly heat-resistant mineral that he called "sappare," now known as kyanite. In addition to being very difficult to melt, sappare could be split into long, thin fragments with pointed ends. Saussure mounted his samples on those ends, often embedding the other end of the fragment into a glass tube used as a holder. To attach the sample to the sappare, he recommended either "saliva or slightly gummy water" or just plain water, but Smithson reported having great difficulty in making any of these adhesives work. "A splinter of sappare," he wrote, "appeared to fulfill the conditions of this problem, and to have accomplished all that could be desired. It has, however, been scarcely at all employed, owing to the excessive difficulty in general of making the particles adhere; and in consequence the almost unpossessed degree of patience required for, and the time consumed by, nearly interminable failures." Smithson's solution to this difficulty furnished the title of this paper.

Smithson had been working on this problem for a long time. As early as 1804, in a letter to the editor of the *Journal de Physique*, he mentioned having substituted a form of flint, "la pierre à fusil," for the troublesome sappare. He may have been referring to the flint used in flintlock rifles, because he noted that it was also called "rifle stone" and was readily available. He made no mention of this mineral in his 1823 article because he had found even better materials and methods. He described replacing the sappare with "small triangles, or slender strips of baked clay," and using "a mixture of water and refractory clay" to attach the tiny sample to them. In a further refinement, he described using the end of a platinum wire as a holder: "almost the least quantity of clay and water is put on the *very end* of a platina wire, filed flat there. With this, the particle of mineral lying on the table can be touched in any part chosen; for a moment or two it is dry, and may be taken up, and put in the flame." Barely visible particles could be tested in this manner, and if there was a concern that using clay might interfere with the test, he described making a paste from the sample's powder and water and using this as the adhesive.

This was what Smithson wanted to convey, and he ended the article with just a few observations about what he had learned with this method. Flint, he reported, melts easily this way while also frothing and giving off a distinctive smell. "Does flint, like pitchstone, contain bitumen," he asked. (It is now known that it often does.) Smithson also described a simple test for distinguishing between a fragment of diamond and one of quartz, which can look almost identical. The solution was to attach a piece of each to one of his small clay strips and test them simultaneously in the flame. He reported that "the diamond was most luminous while under the action of the flame, and longer so after removal from it." A method of comparing two different substances in the same blowpipe flame had not been previously described.

Not surprisingly, the reaction to this article was confined mostly to technical publications, but for the many users of the blowpipe, Smithson's new methods were of considerable interest. His entire article was reprinted in the *Technical Repository* (1824), and summaries of it appeared in journals such as the *Quarterly Journal of Science* (1823), the *Bulletin des Sciences* (1824), and the *Annales des Mines* (1826). An article praising Smithson and expanding on his techniques was read to the newly founded Lyceum in New York and quickly published in its journal.

The greatest impact probably came from the inclusion of Smithson's methods in nineteenth-century blowpipe manuals. Berzelius, for example, in his *Use of the Blowpipe in Chemistry and Mineralogy* (1845), characterized Smithson's use of clay strips and platinum wires as common methods of support, and he credited Smithson for having made "important additions" to the use of the blowpipe. Michael Faraday described how to use Smithson's clay strips in his textbook *Chemical Manipulation* (1842).

One of the characteristics of Smithson's science was his almost relentless effort to improve the tools and methodology of his work. In this case, his efforts to improve on the sappare spanned nearly two decades, and even after that, he continued to look for new methods. He mentioned in his article that he had "only recently" begun to use the end of the platinum wire as a method of support. These developments consistently allowed him to work with smaller and smaller samples, which was in keeping with his role as one of the pioneers of what would come to be known as microchemistry.

16.

AN "ARISTOCRATIC
SCIENCE DABBLER"?

IN 1823 James Smithson wrote two articles that have proved problematic for his biographers. One was about the shape of ice crystals, the other about a method he had developed for making coffee while traveling. More than a century and a half after Smithson wrote them, these articles—especially the one on coffee—were presented as evidence that "Smithson's career was marginal, maybe even irrelevant" and that Smithson himself had been an "aristocratic science dabbler." The articles have always shackled Smithson's scientific reputation, and to a modern reader they do seem unscientific. Was this the context in which he wrote them?

The first article was a short letter to his friend Thomas Thomson, editor of *Annals of Philosophy*. Smithson had just finished reading about large ice formations that existed year-round in certain underground caves and had been struck by the description of one in southern France. It contained large hollow ice columns filled with beautiful ice crystals and a thick ice floor "entirely composed of crystallized parts presenting hexahedral prisms." Smithson had never written about ice crystals before, but his reference to the work of seven other writers in the brief introduction showed his familiarity with the topic's scientific literature.

This interest in ice may seem surprising, since it falls outside of Smithson's focus in chemistry and mineralogy. But he was also deeply interested in crystallography, and the study of ice crystals was well established in that discipline. As is the case with most minerals, when water moves from a liquid to a solid state it generally forms a mass without any particular shape. Under the right conditions, also like many minerals, water can solidify as regular crystals. This is most apparent in snowflakes, which are all different but always have a fundamentally hexagonal shape—something that had been of scientific interest since the early seventeenth century. Johannes Kepler noticed it, and in 1611 he wondered "if it happened by chance, why would they [snow] always fall with six corners and not with five, or seven?" In 1635, René Descartes described finding snow in the form of "little plates of ice, very flat, very polished, very transparent, about the thickness of a sheet of rather thick paper . . . perfectly formed in hexagons."

There wasn't much doubt about snowflakes being six-sided, but the same could not be said about the crystals sometimes seen in ice. Ice crystals, like those of many minerals, are often jumbled together and hard to distinguish. The shape of ice crystals was still disputed when Smithson wrote his article, and he pointed out that well-respected authors had reported them to be either hexagonal, triangular, rhombic, or octagonal. "Are these accounts and opinions accurate?" he asked.

Part of the problem was that ice crystals are usually short-lived. For example, Edward Clarke of Cambridge once reported finding rhomboid-shaped ice crystals growing on the timbers of a wooden bridge. Clarke was a knowledgeable mineralogist, but the crystals he found only lasted three days—which made confirming his observation problematic, especially when no other rhombic ice crystals had been reported. This was why Smithson found the article about ice caves so interesting: the crystals in these caves existed year-round.

Smithson did not address the claims of others, but instead reported his own observations of a form of crystalline ice that was widely available for examination: hail. The crystalline form was not always apparent in hailstones, but when it was, he wrote, "It is constantly that of two hexagonal pyramids joined base to base." The processes that form hail are variable and extremely complex, and Smithson did not suggest that all hail had this form. But in those cases where it could be discerned, he found this to be the crystal's shape. Certainly, his long study and experience in crystallography helped him in this determination, for it is unusual to find hail in the form of a single, whole crystal. Indeed, Smithson reported that "I do not think that I have measured the inclination of the faces more than once. The two pyramids appeared to form by their junction an angle of about 80 degrees." This was apparently the only angle measurement he was able to make.

Smithson also reported another feature about ice crystals, and this too drew on his experience with crystallography. He observed that in the crystals found in hail, "one of the pyramids is truncated, which leads to the idea that ice becomes electrified on a variation of its temperature." In mineralogy, when one end of a crystal is different from the other, it often means that the crystal will develop electric charges on those ends when heated. Smithson had made a study of these "electric" crystals as a young man, and he mentioned tourmaline and "silicate of zinc" as examples of crystals with this property.

Smithson's description of the crystals in hail and his suggestion that they are "electric" correspond closely to the modern understanding. Virtually all the ice found on the Earth's surface is now seen as some form of hexagonal crystal, either as plates, columns, or the six-sided (sometimes truncated) bipyramids described by Smithson. Hail is now known to be electrified prior to reaching the ground by collisions with particles having different temperatures.

Smithson also made an interesting observation about snow. Snowflakes, he wrote, are "the same form as hail, but imperfect. Its flakes are skeletons of the crystals, having the greatest analogy to certain crystals of alum, white sulphuret of iron, etc., whose faces are wanting, and which consist of edges only." This was a reference to the fact that in many kinds of mineral crystals, the edges and corners grow the fastest, and can sometimes grow over undeveloped interior spaces. Smithson suggested a similar mechanism to explain the appearance (and variations) in snowflakes. It was an intriguing idea, although this is not the modern understanding. Today, snowflakes are still considered to be fundamentally crystalline, which explains their six-sided symmetry, but their exact shape is determined by the specific temperature and humidity conditions they pass through as they fall.

Despite its short length, Smithson's article was widely noticed. His identification of the shape of ice crystals was repeatedly cited throughout the nineteenth century, particularly in German texts, and there were repeated references to Smithson's article in *Neues Jahrbuch für Mineralogie* (1886). He was also credited with the identification of pyramidal ice crystals in French and English texts, such as Gabriel Delafosse's *Nouveau cours de minéralogie* (1862) and James Fowler's *Water: Its Nature and Natural Varieties* (1865). While Smithson's work on ice may not have been a seminal contribution to science, it was seen as important at the time, and he was widely credited for it.

Yet this article was barely mentioned when the Smithsonian Institution published his scientific writings in 1879, even though meteorology had long been a topic of special interest there. Part of the reason may have been that many of the references to Smithson appeared in German scientific publications, which were not searched when the institution appraised his work. This may also be another case of Smithson's reviewers simply failing to appreciate the breadth of his interests and his work. As with his articles on geology, plant chemistry, and electricity, Smithson's article on ice did not fit into their understanding of him as exclusively a mineralogist and inorganic chemist, and so it was dismissed as his indulging in an obscure topic.

The other article that has seemed dilettantish to some Smithson biographers is "An Improved Method of Making Coffee." Most have believed that coffee making is not a proper scientific topic, one going so far as to say that Smithson "liked to tinker with coffee." But that judgment fails to appreciate the long and special relationship between coffee and English science.

Coffeehouses began appearing in England in the 1650s, toward the end of the English Civil War. They offered a welcome alternative to taverns and alehouses and almost from the beginning became a gathering place for those interested in natural philosophy. The origins of the Royal Society, England's oldest and most venerable scientific institution, can be traced to these early coffeehouses, and its members continued to frequent them throughout the eighteenth century.

By 1700 London was said to have more than two thousand coffeehouses. Most were open to anyone who could pay the penny admission, but some were noted for attracting a more specialized clientele. One of these was the Grecian coffeehouse in Devereux Court, a famous haunt of scholars and savants. Isaac Newton and other members of the Royal Society often went there after the society's regular Thursday meetings concluded, and this was where Edmond Halley, the royal astronomer, stopped to meet friends on his weekly visit from Oxford to London. Sir Hans Sloan, whose collection of curiosities would later be used to establish the British Museum, was also a Grecian regular.

Newton's group was popularly known as the "Learned Club," and throughout the eighteenth century it was common to find similar "clubs" at the favorite coffeehouses of important scientific figures. Martin Folkes, president of the Royal Society after Newton, headed such a club, which met at the Baptist Head coffeehouse in Chancery Lane. While he lived in London in the 1760s and 1770s, Benjamin Franklin (who, in addition to his other accomplishments, was a noted natural philosopher) headed the "Club of Thirteen," which met at Old Slaughter's coffeehouse. Joseph Banks, who was president of the Royal Society in Smithson's time, attended a scientific club that met at Jack's coffeehouse. Much more than social gatherings, these clubs were where the regular business of science was conducted. Information exchange, identification of new questions, scientific discussion and debate, and even the occasional experiment, all took place in London coffeehouses.

Although they met in public houses, membership in these elite clubs was by invitation only. However, some of the less prestigious scientific clubs were more welcoming. One of these was the Society for Promoting Natural History, which Smithson joined while he was still in college. Distinguished by its network of foreign correspondents, the club provided a way for those who (like Smithson) were new to London to join the scientific community.

In the spring of 1786, Smithson was elected to a more prestigious club, the "Coffee House Philosophical Society," which at that time was meeting at the Baptist Head coffeehouse. Organized by the Irish chemist Richard Kirwan, who invited him to join, the club became an important way for Smithson to meet and impress new people. This seems to be what happened, because the following year he was elected to the Royal Society.

This intermingling of science and coffeehouse culture was more pronounced in England than elsewhere in Europe, and the English were happy to acknowledge coffee's beneficial effects. As Benjamin Thompson, one of the founders of the Royal Institution, famously remarked, "It has been facetiously observed that there is more wit in Europe since the use of coffee has become general among us; and I do not hesitate to confess that I am seriously of that opinion. Some of the ablest, most brilliant, and most indefatigable men I have been acquainted with have been remarkable for their fondness for coffee." Others praised coffee's ability to suspend "the inclination to sleep" and noted that coffee was "particularly adapted for studious and sedentary people." But for all its stimulating properties, it is not clear whether English coffee actually tasted very good.

The standard method of making coffee in English coffeehouses seems to have remained unchanged from the end of the seventeenth century to the beginning of the nineteenth century. The beans were roasted, pounded into a powder in a mortise, thrown into a kettle with a quantity of water, and then boiled. After a while the kettle would be removed from the fire, allowed to sit a few minutes while the grounds settled to the bottom, and then served. If there were coffee grounds floating in it, as was frequently the case, most could be settled by adding a few spoons of cold water. The coffee was served in bowls and ideally would be drunk very hot, because the desirable parts of the coffee were thought to exist in the vapors and were quickly lost as the coffee cooled.

Around Smithson's time, the practice of adding cold water to settle the grounds was replaced by adding isinglass, a material made from the air bladders of fish. Isinglass had long been used as a "fining" or clarifying agent in the making of beer and wine, and it now performed that same action in coffee by consolidating the floating grounds into a jelly-like mass that sank to the bottom of the kettle. This eliminated the wait for the grounds to settle and allowed the coffee to arrive at the table—and be consumed—that much hotter.

As far as the strength of coffeehouse coffee, that seems to have varied. Most accounts describe using one ounce of coffee per quart of water, but tastes varied. The writer of a 1722 treatise opined, "I cannot conceive how you can put less than two ounces of powder to a quart, or one ounce to a pint of water; some put two ounces and a quarter."

What that meant in terms of the coffee's taste is interesting to consider. If the typical ratio is assumed to be one ounce coffee to one quart water, then they were using about half a tablespoon of coffee per cup, which is light by today's standards. But since the coffeehouses' coffee was powdered and brewed significantly longer, they extracted more of the coffee bean's contents, both good and bad, which would have made it strong and quite bitter. Add to this the fact that their coffee was made with water taken directly from the Thames River, and it is not hard to see why they preferred to drink it quickly and very hot.

In chemical terms, the coffeehouse method of extracting coffee with boiling water was a "decoction," but there were also other extraction methods, and one of these was the "infusion." The first infusion coffee pot appeared in France in 1711 and consisted simply of a cloth bag that held the coffee inside the pot while soaking in the hot water. Similar to a teabag, it had the obvious advantage of producing coffee without floating grounds. It also eliminated the step of boiling the coffee, which presumably improved the taste. By 1760 the infusion, or "steeping" method, had largely replaced the decoction method in France, but it failed to catch on in England.

Another method of extraction was "percolation," which appeared in France in the early 1800s. Using this method, coffee is made by pouring boiling water over powdered beans held in a perforated basket. Since the water was only in contact with the beans for a short time, percolation extracted only the most desirable compounds. The great English advocate of percolation was Sir Benjamin Thompson, better known as Count Rumford, who was an enthusiastic coffee drinker. He and Smithson were both members of the Royal Society and the Royal Institution and may have been acquainted. Rumford was a major scientific figure, and like many Enlightenment natural philosophers, including Smithson, he was interested in applying science to the public good. He made a long, detailed study of coffee in which he applied the principles of experimental philosophy to each step of the coffee-brewing process. The resulting illustrated monograph, published in 1812, was more than fifty pages long.

Since his goal was to make the perfect cup of coffee, Rumford first needed to define his terms. He surveyed the capacity of English and French coffee cups, which he discovered typically held 8 1/3 cubic inches of coffee. This was close to the

English gill, a common English measure at the time, equivalent to about half a cup today. Rumford decided to use this as his standard unit of coffee. Having determined the standard cup size, Rumford next set about to determine the precise amount of coffee necessary "to make a gill of most excellent coffee, of the highest possible flavor and quite strong enough to be agreeable," but he quickly found this to be critically dependent on the method by which the coffee was made. Rumford spent the next fifteen years studying, experimenting, and improving each step of the coffee-brewing process.

He developed a method of roasting coffee in sealed glass containers, which gave him more control over the roasting process and helped to retain the important vapors. He designed a storage container for roasted beans, which had a sliding piston inside that pushed down on the beans and sealed them off from the air. He also performed extensive experiments to find the best water temperature, setting up taste tests where he made multiple batches of coffee in identical pots, with the heat of the water as the only variable. Eventually he determined that the weight of powdered coffee needed for the perfect cup was "108 grains Troy" which, for those outside the scientific community, he rounded off to a quarter of an ounce—or half a tablespoon.

Rumford also used experiments to design a special percolator, optimized to make four cups of perfect coffee. It had a specially designed strainer to hold the coffee and a rammer to level and compress the coffee to a uniform thickness. This rammer was left in place during brewing so that the coffee was not displaced when the boiling water was added. Because of this design, it took between eight and ten minutes for water to percolate through the coffee, which meant that it needed to be kept warm. Rumford designed a special oil lamp for the percolator to sit on and gave the pot itself a double-wall construction that could be filled with boiling water at the start of the brewing process.

Rumford did not invent the percolating coffee pot, and he did not claim to, but his article was so influential that it has frequently been attributed to him. Although he contributed to the popularity of "percolators" for home use, very few pots using his double-wall design were manufactured. Most pots of this style used a less expensive single-wall construction and, as a consequence, had a tendency to produce coffee that was no longer hot. Smithson referred to this problem in the beginning of his article, noting that coffee made by "mere percolation" was "apt to be cold." The brewing method he recommended instead was one he had developed during his years of travel. First combine water and powdered coffee in a bottle, seal it with a cork, and then heat it in a "water-bath"—a pan of boiling water. This simple method had the advantage of extracting the coffee gently and preserving all the fragrant vapors, which were lost with other methods.

When the coffee was ready, Smithson poured it through a paper filter to remove the grounds. At that point it could be drunk or put back in a bottle and returned to the water-bath for later consumption. As long as the bottle was sealed, it would not lose its flavor. He reported that coffee made this way "may be kept for any length of time at a boiling heat, in private families, coffee houses, etc., so as to be ready at the very instant called for." He also noted that the method was convenient for travelers, since it eliminated the need to carry a bulky kettle or coffee pot. All they needed was a sealable bottle and something to boil water in.

Smithson's method is technically a filtered infusion, and as such it differs slightly from other coffee-brewing methods of the time. Because none of the aromatic vapors escape and boiling water never actually touches the coffee, it produces a smooth, full-flavored coffee that is not bitter and may not need the addition of cream or sugar. Smithson suggested other possible uses for this method of extraction, including brewing beer and preparing medicines. But he clearly saw its greatest utility in the economical preparation of coffee, "which constitute[s] one of the daily meals of a large portion of the population of the earth." Unfortunately, Smithson never made any effort to develop his method into a convenient system for regular use, and it was quickly forgotten.

Rumford and Smithson were not the only natural philosophers interested in coffee. Coffee was an important topic throughout Smithson's lifetime, and across Europe (and especially in France) chemists and physicians were making a determined push to isolate coffee's active ingredient. This was finally achieved in 1819 with the discovery of caffeine. Rumford and Smithson were unique in their interest in improving the preparation of coffee, an interest that may best be seen as growing out of their Enlightenment values. In their writings on coffee, Rumford and Smithson both spoke of the purpose of science being to improve the lot of the general population. Smithson was a wealthy aristocrat, but he justified studying coffee preparation because "in all cases means of economy tend to augment and diffuse comforts and happiness. They bring within the reach of the many what wasteful proceedings confine to the few." In the decades after Smithson died, concern for everyday comforts would increasingly be seen as lying outside the scientific realm, but in Smithson's time it was not. In writing about making coffee, he was adding to a well-established scientific topic.

———————— + ————————

17.

CHLORIDE OF POTASSIUM

SMITHSON PUBLISHED SIX ARTICLES IN 1823. Only a single page long, the fifth described his analysis of a sample brought to him from Mount Vesuvius. He reported that it was "part of a block which was said to have been thrown out" of the volcano during a recent eruption. He called the article "A Discovery of Chloride of Potassium in the Earth."

Smithson described the sample as "a spongy lava," consisting of a "red ferruginous mass" with a few crystals imbedded in it "here or there," and with veins of a white crystalline matter "more or less disseminated through nearly the whole of the mass." The red mass he judged to be ordinary lava colored red by the presence of iron oxides, and the crystals he identified as either augite or hornblende, which commonly form in lavas as they cool. These materials were unremarkable and of little interest, but the white crystalline material that filled most of the sample's spaces was a different case. He reported that through a lens it had a "saline appearance," a "tabular fracture," and, in a few places, "regular cubical crystals." Basing his conclusion strictly on this physical examination and drawing from his extensive experience, Smithson supposed it to be either ordinary table salt (NaCl), or "muriate of ammonia" (NH_4Cl). These assumptions became the starting point for his analysis.

That analysis consisted of just six experiments that Smithson described in a total of eight sentences. It was a methodical and strictly qualitative investigation. Nothing was weighed or measured. Smithson described a series of tests in which he manipulated the unknown material with heat or other chemicals. Each test either confirmed or ruled out one of his assumptions and moved him one step closer to identifying the unknown material. As with so many of his articles, there was no discussion of what any of the individual tests revealed, but by returning to what was known at the time, we can follow the path of his reasoning.

In the first test he heated the unknown material in a "matrass," a tall cylindrical glass vessel that made it easier to see if any gases were being released. Muriate of ammonia, one of the two suspected materials, was known to produce a white smoke before it melted. Although he heated the sample until it melted, he reported that it produced "little or nothing" in the way of gases, ruling out muriate of ammonia as the unknown salt.

In the second test Smithson explored the other possibility: that the unknown material was ordinary table salt. He reported that it "dissolved entirely in water" as would be expected, but when "laid on silver with sulphate of copper, it produced an intense black stain." Smithson was describing a test for sulfur, one that he had used before. It made use of the simple fact that if a solution containing sulfur was placed on a piece of silver, it would make a dark stain. It was a simple and surprisingly sensitive test, but there were situations where it would not work, and one of them was if the solution contained table salt. The salt would keep the black stain from forming. So, if the solution he tested contained table salt, this test should have produced no reaction even though sulfur was present. But Smithson reported that exactly the opposite happened, that the solution produced "an intense black stain," which meant that the unknown material was not table salt.

In the third test, just to be sure that the unknown material did not contain sulfur, he tried adding "chloride of barium" to a solution of the unknown salt and water. Barium was known to have a strong attraction to sulfur and would quickly bond with it to form a thick white precipitate on the bottom of the container. Smithson reported seeing no such reaction—only a "very slight turbidness" in the water, which he attributed to impurities in his sample. The unknown substance did not contain sulfur.

Having eliminated the most likely possibilities, Smithson drew on his experience to try a hunch. Something about the new mineral had given him an idea, and in the next test he added a completely different type of material to the salt solution, "tartaric acid" ($C_4H_6O_6$). This organic acid, Smithson reported, "occasioned an abundant formation of crystals of tartar."

"Crystals of tartar" is another term for bitartrate of potash, which in Smithson's time was used in cooking, especially in meringues, and sometimes medically as a purgative. It was produced by adding tartaric acid to a solution of chloride of potassium and water. Adding tartaric acid to the unknown solution produced these same white crystals, identifying the unknown material as chloride of potassium (KCl).

This was the first instance of this substance being found in nature in its pure form, so Smithson performed a fifth and sixth test, just to be sure. In the fifth he added chloride of platinum to the salt solution and reported that it "immediately threw down a precipitate, and distinct octahedral crystals of the same nature afterwards appeared." These distinctive yellow crystals appeared to be "platinic-potassium chloride," a substance known to be produced by adding chloride of platinum to a solution of chloride of potassium, which seemed to confirm Smithson's analysis.

In the sixth and final test, Smithson added nitric acid to the salt solution, which formed long, narrow crystals that he identified from their appearance as "nitrate of potash." Because nitrate of potash was known to be the product of nitric acid and chloride of potassium (in solution) this was all the confirmation he needed. "It appears from these experiments, that this white saline matter is pure, or nearly pure, chloride of potassium," he wrote, attributing its presence on the lava sample to "sublimation"—a high-heat phenomenon where a solid changes directly into a gas. Smithson appears to have known that potassium chloride has a very high boiling point (almost 2600°F, higher than the temperatures found in most volcanoes) and he was using that to suggest a method by which this deposit could have formed. The sample was said to have been "thrown out" of the volcano, and it may have been accompanied by some of the potassium chloride gas, which immediately condensed on it after being ejected from the volcano.

Smithson's finding was new for both mineralogy and volcanology, but it does not seem to have attracted much interest. The only English notice of this article was in a short note in the *Journal* of the Royal Institution, and French journals do not seem to have commented on it at all. Three German scientific journals mentioned it, but after that it was forgotten. Smithson was never credited for the discovery of the mineral, and in 1832 it was named "sylvite," after a Dutch physician. Some of the early editions of James Dana's popular *System of Mineralogy* did credit Smithson for the suggestion that sylvite (or syline, as it was sometimes called) formed at the fumaroles of volcanoes, but by the 1884 edition his name had been separated from that insight.

If Smithson wanted to generate more interest in his article, he needed only to comment on its implications for some of the scientific theories then being discussed, such as Humphry Davy's "chloridic theory." This theory, which was still being actively debated at the time, had grown out of Davy's identification of the element chlorine. In explaining the properties of the various "chlorides," particularly the solutions they formed when dissolved in water, Davy had moved toward a new understanding of the chemical bond. Smithson's use of such solutions in his analysis gave him several opportunities to comment on this theory, but he chose to remain silent.

Neither did Smithson comment on the significance of finding another potassium compound among the products of volcanoes. It had been ten years since he wrote the controversial "Saline Substance" article supporting Davy's theory of volcanoes. That theory was still being debated, but Smithson confined himself to a straightforward analysis of the sample and resisted the temptation to speculate on its broader implications. As a young man he had been known for his deep interest in the philosophical side of science, but he was fifty-eight years old now and not as inclined to speculate in his scientific papers.

18.

COMPOUNDS OF FLUORINE

ON THE SECOND DAY of January 1824, James Smithson sent off an article that he had been thinking about for a long time. It was titled "On Some Compounds of Fluorine," but it was primarily about one compound, the mineral "fluor spar," better known today as "fluorite" (CaF_2), a common mineral that comes in a surprising range of attractive colors. As Smithson pointed out, "Fluor spar has decorated mineral cabinets [mineral collections] from probably the earliest period of their existence." Even today, only quartz is more commonly collected. The earliest written references to fluor spar date from the sixteenth-century, but it was used long before that as a flux in the production of metals, helping the metal flow out as the ore was heated. Its curious name, "fluor," derived from the Latinized *fluere*, meaning "to flow."

The Swedish chemist Karl Scheele had analyzed fluor spar in 1771. He reported that when placed in sulfuric acid, it decomposed into calcium and a powerful new acid with some surprising properties. One was that it dissolved glass, and in its gaseous form the acid had actually eaten holes in some of Scheele's glass apparatus. The second was that when it came in contact with water, it formed a mineral-like crust that resembled quartz. The third, which does not seem to have been immediately appreciated, was that it was extremely dangerous to work with.

Word of Scheele's discovery generated great interest within the scientific community. Within a year Joseph Black, the great Scottish chemist, was performing his own experiments and pressing his friend James Watt, who was also interested, to send him fluor spar samples from Cornwall. The English chemist Joseph Priestley also experimented with "fluor acid," as he called it, heating it to study its gaseous state. Heating, he reported, greatly increased the acid's corrosive property, and the thickest glass apparatus he could find rarely survived more than an hour's exposure before the acid ate holes in them.

Scheele, Black, and Priestley were three of the most accomplished chemists of their time, but none of them were able to properly describe the composition of the new acid. Scheele was incorrectly convinced that it contained silica. Black's main contribution was to find new uses for it as a flux for metals. Priestley, as an advocate of the soon-to-be-obsolete "phlogiston" theory, identified it as simply a form of sulfuric acid "charged with so much phlogiston as is necessary to it taking the form of air."

Smithson was just six years old when Scheele's analysis was published, and although other chemists continued to study it, the nature of the acid was still a mystery when he attended college. It was mentioned in his chemistry class at Oxford, where it was called "Mr. Scheele's Acid of Spar," but only Scheele's experiments were discussed. There was something in fluor spar, and in the acid it produced, that had yet to be identified and seemed to be unique. In 1789, in his revolutionary *Elements of Chemistry*, Antoine Lavoisier acknowledged this by listing the mysterious substance as one of the chemical elements; even though it had not yet been isolated, he named it "fluoric."

Isolating this new element would become one of the fundamental challenges of chemistry for the next century. Among the few surviving notes from Smithson's mineral catalogue is that in 1799 he acquired samples of fluor spar "from Matlock in Derbyshire." Although fluor spar deposits were widespread in England, most of the mineral studied by chemists came from just a few mines in a small area of Derbyshire, near the town of Matlock. Known as "Blue John" by miners and "Derbyshire Spar" by chemists and collectors, this variety was distinctively banded white, yellow, and purple. Priestley reported using it, as did the translator of Lavoisier's *Elements*. Later, in their experiments with the acid, Humphry Davy and Thomas Thomson also specifically used it. The fact that Smithson acquired specimens from this particular location is strong evidence that he was preparing to perform experiments of his own.

Eighteenth-century chemical analysis was able to identify only the materials with which this new element combined. But in the early nineteenth century, the successful use of electricity to discover other new elements inspired a new round of attempts to isolate what was coming to be called "fluorine." In 1810, Humphry Davy predicted that a new element would be found in fluor spar and asked his brother, also a gifted chemist, to do some preliminary research. In 1813 Davy published the results of his first study of compounds containing fluorine, quickly followed in 1814 by a second article on the same topic.

Smithson was living in London at this time, and he and Davy were colleagues at both the Royal Institution and the Royal Society, so it is not surprising to see an overlap in their interests. At the end of his 1811 article on zeolite, Smithson included an aside suggesting that the mineral topaz ($Al_2SiO_4(FOH)_2$) was "a compound salt, consisting of silicate of alumina, and fluate of alumina." In response, Davy wrote in his 1813 article that "new experiments are required to shew whether that gem is a true silicate fluate of alumina, or a compound of the inflammable bases of alumina and silica with fluorine." More than a decade later, in "Compounds of Fluorine," Smithson revealed that he had been thinking about this question and that his opinion had changed. "I am now convinced," he wrote, "that no oxygen exists in it; but that it is a combination of the fluorides of silicium and aluminum." In declaring that topaz contained no oxygen, Smithson may have been responding to Davy's suggestion that fluoric acid was a combination of fluoride and hydrogen, which later proved to be the case. Smithson, however, was wrong that topaz contains no oxygen. Later in his "Compounds" article, Smithson revealed that he had also experimented with the mineral "kryolite" (Na_3AlF_6), although without any new discoveries. This mineral contains fluorine, and Davy had earlier identified it as needing further investigation.

The question of the composition of the fluoric compounds, and specifically whether fluorine combined with oxygen or hydrogen, was related to the fundamental nature of acids, which was one of the great questions of the age. In continuing to work on fluorine, Smithson joined some of the most distinguished chemists of his time: Humphry Davy, John Davy, André-Marie Ampère, Louis Joseph Gay-Lussac, Thomas Thomson, and Jacob Berzelius all spent significant amounts of time researching this question, and Smithson knew all of them. Because fluorine is so reactive, they tried to contain it in apparatus made from materials that did not readily combine with it. Davy used small platinum vessels and experimented with tubes made of sulfur. Gay-Lussac and his associates used vessels made from either silver or lead. But fluoric acid gas is an acute poison, and even a brief exposure can permanently damage the lungs and corneas, while the liquid form can cause deep burns and tissue death. As research continued into the late nineteenth century, the number of chemists injured or killed grew ever larger, and they came to be known as the "fluorine martyrs." That term was coined after Smithson's time, but at least two of his friends were on that list. Humphry Davy and Gay-Lussac were both injured in accidents with fluorine, and although they eventually recovered, Smithson was clearly sensitive to the danger of working with this strange substance.

Before Smithson's "Compounds" article, there was only one general test for detecting fluorine in chemical analysis, and it took advantage of fluoric acid's ability to dissolve glass. A drop or two of sulfuric acid was placed on a polished glass plate, and a tiny piece of the sample was placed in it and gently heated. If the sample contained fluorine, fluoric acid would be produced and would faintly etch the polish from that part of the glass underneath the drop. There were other, confirmatory, tests for specific compounds once fluorine was suspected, but this was the only general test. It was not ideal. One text advised, "A barely perceptible etching is made more visible by breathing upon the glass," but warned that "if much silicic acid is present, this reaction fails." What was needed was an unambiguous and safe test for the "fluoric element," as Smithson called it, and in this article he presented one.

Smithson described a blowpipe test built upon a suggestion made by the Swedish chemist Jacob Berzelius. Berzelius had recommended using a glass tube to hold materials being heated. As the sample is heated, he wrote, "The volatile substances, not permanently gaseous, sublime and condense in the upper part of the tube, where their nature may be ascertained." It was an effective technique, but one of the disadvantages of the tube was that the glass could quickly melt if a high heat was applied. Smithson's modification was to place either a piece of platinum foil or a small curved piece of baked clay at one end. This served both to hold the sample and protect the glass from melting. As long as the tube was held at an angle, the force of the blowpipe flame would direct any materials released into it, where they could be detected either by residues they left inside or, in the case of gases, by color changes to indicator papers placed in the far end of the tube. Smithson detected fluoric acid by the combined effect it had on the inside of the tube and on a small piece of indicator paper inserted into the top. For fluorine compounds, Berzelius recommended paper tinged with the reddish juice of "fernambuc wood" (Brazilwood), which, Smithson reported, is made yellow by fluoric acid and oxalic acid, but not by sulphuric, muriatic, or phosphoric acid. Since oxalic acid did not etch glass, any substance that turned Brazilwood paper yellow and etched the inside of the glass tube was identified as fluoric acid gas. Smithson reported that papers colored with the blue juice of logwood also served the same purpose.

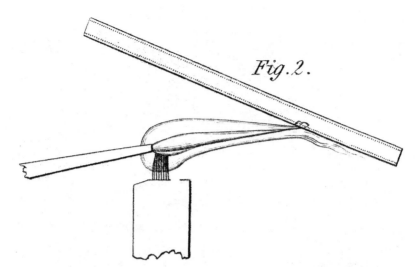

Berzelius's drawing of a blowpipe (left) being used with a lamp flame (bottom center) and a glass tube (upper right). The flame is being directed at the sample, which is inside the tube. Any gases released by the sample would naturally flow up the hollow tube, where they could be studied. Courtesy of the Smithsonian Libraries.

This apparatus could be used to test any material for fluorine, but in what may have been an abundance of caution, Smithson also devised a special apparatus for "substances of difficult fusion." Although challenging to construct, it simplified holding materials like topaz, which needed a long exposure to the flame to dissociate its fluoric part. It was intended to be used with extremely small samples, even powders, which were mounted on the end of a platinum wire "nearly as fine as a hair." The opening of the tube was covered by platinum foil to protect it.

The apparatus Smithson described extended the range and utility of the blowpipe as an analytical tool, and this was arguably the most important contribution of his article. A summary appeared in the French *Bulletin des Sciences Mathématiques*, and some of Smithson's observations about fluor spar were repeated in Gmelin's *Hand-Book of Chemistry* (1849), but most other references to the article appeared in blowpipe manuals. Berzelius discussed Smithson's modification of his "tube de verre" (glass tube) in his book *Traité de chimie* (1831), and he recommended it in *The Use of the Blowpipe* (1845), which was also translated into German. The specialized apparatus does not appear to have been widely used, but Smithson's improvement to the glass tube—placing a clay or platinum holder into the tube's end—became a standard method. As late as 1885, the venerable *Plattner's Manual* of blowpipe analysis still recommended it for testing minerals for

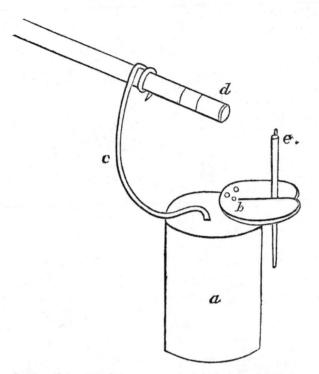

Smithson's drawing of his apparatus for "substances of difficult fusion." Because fluoric acid is so hazardous, he designed it to work with extremely small samples. The sample, shown here as "e," was attached to the end of a fine platinum wire with clay, a technique that he developed and had written about in an earlier article. As an indication of the size of this apparatus, "a" is the cork from a wine bottle. In use, the apparatus was held in one hand and positioned so that the blowpipe flame would drive any gases from the sample into the glass tube. Courtesy of the Smithsonian Libraries.

the presence of fluorine, asserting that it could detect the gas "when only three-fourth per cent of fluorine is present." Smithson's apparatus finally faded from use in the early twentieth century, as blowpipe analysis was replaced by other, more quantitative, tests.

Smithson's article also contained observations about working with fluor spar, and he included some fairly detailed suggestions about experiments that might lead to new insights into the nature of fluorine. But, he remarked, "It is not convenient to myself at present to make the experiment: I therefore resign it to others." And here might have been Smithson's reason for writing this article. Now nearly sixty and in failing health, Smithson seems to have realized that the time for him to work on this long-standing question had come to an end, and others would have to carry it forward. There is a sense in this article that Smithson is trying to encourage interest in the topic, and he presents his blowpipe tests like a gift to future researchers.

For those who might object that the tiny samples he used and the ephemeral effects they produced were hardly likely to lead to a new understanding, he offered one final thought. "There may be persons," he wrote, "who, measuring the importance of the subject by the magnitude of the objects, will cast a supercilious look on this discussion; but the particle and the planet are subject to the same laws; and what is learned upon the one will be known of the other."

———————+———————

19.

EGYPTIAN COLORS

ON JANUARY 2, 1824, the same day he sent his publisher the "Compounds of Fluorine" article, Smithson also sent a second article. This one was titled "An Examination of Some Egyptian Colors," and it described his analysis of paint pigments used by ancient Egyptians. While this may seem like a curious topic, it was, like fluorine, something that had interested him since he was a young man. Indeed, it had long been a topic of interest to many natural philosophers. Isaac Newton studied and speculated about them, as had Martin Wall, Smithson's chemistry teacher at Oxford. In his inaugural address, given just one year before Smithson enrolled, one of the topics Wall discussed was "the proverbial celebrity of the Egyptians in general science, and in many of the arts."

Napoleon's 1798 military expedition to Egypt dramatically increased this interest in Egypt. In addition to his army, he took teams of scientists and scholars with him, and the artifacts, artwork, drawings, and reports they sent back to France created a sensation. This fascination with Egypt spread to England a few years later, fueled by the wave of English tourists who rushed to visit as soon as Napoleon was expelled. Smithson's half-brother, Henry Louis Dickenson, was among those early tourists, and in 1803 he brought Smithson a gift, a small figure of the Egyptian goddess Isis. The figure was painted blue, and when Smithson analyzed a bit of the pigment, he found it to be a kind of glass colored with copper by an unknown process. He chose not to publish this finding at the time, probably because the age and origin of the artifact were unknown, but he would return to the topic many years later.

England's fascination with all things Egyptian continued throughout the first quarter of the nineteenth century, and around 1815 it reached a fevered peak. Under Napoleon, the French had sent impressive quantities of Egyptian relics back to Paris, and now, with Napoleon defeated and the long war finally over, the English felt it was their turn. Henry Salt was the newly appointed British Consul-General in Egypt, and among his mandates was one from the trustees of the British Museum "to collect Egyptian antiquities for our great national depository." To carry this out he chose one of the most colorful figures of the age, Giovanni Battista Belzoni.

Born the son of a poor Paduan barber, Belzoni showed an early aptitude for engineering, but as the turmoil from the French Revolution spread across Europe those plans were shattered, and he was forced to spend his early years as a sharp-witted itinerant merchant. In his early twenties Belzoni found himself in England where, with no better prospects, he joined the troupe of a small London theater. There he played parts like the giant in *Jack the Giant Killer* and performed feats of strength as the "Patagonian Sampson." These roles were obvious choices consider-ing he was probably around 6'7"—extraordinarily tall for the time—with massive shoulders and a barrel chest.

Belzoni spent more than eight years in the theater, touring the British Isles with several companies and developing his skills as both a showman and pro-moter. However, he seems to have longed for employment that would display his other abilities as well, and in 1815 he went to Egypt, where he hoped to secure employment as an engineer. That commission failed to materialize, but Belzoni had arrived in Egypt at a unique historical moment, a time that has been likened to the gold rush in nineteenth-century America. Egyptian artifacts and monu-ments would fetch handsome prices if they could be found and somehow brought to England, and the British consul counted on Belzoni to make this happen.

Belzoni's first commission was to transport the colossal red-granite head of Ramses II from Thebes to the British Museum. The head weighed at least seven tons and moving it across the soft Egyptian sand and onto a ship was a daunting task, but Belzoni succeeded and that led to a series of other commissions. Boat-loads of Egyptian antiquities that he collected soon began to arrive in London, and the demand for more seemed insatiable. To meet that demand, Belzoni began to conduct his own excavations, and during the next three years, he uncovered the buried temple of Abu Simbel, discovered six new tombs in the Valley of the Kings, opened the Second Pyramid, and found the lost city of Berenice.

The most spectacular of these finds was the tomb of King Psammis, the great Nineteenth Dynasty pharaoh now known as Seti I. Still in relatively undisturbed condition, Tomb KV 17, as it is now known, was the largest and most spectacular tomb yet discovered. In what today would be considered an outrage, but which raised hardly an eyebrow at the time, Belzoni removed everything he considered to be of value and shipped it all back to England. In 1820 he returned to England where, as the tomb's discoverer, he was generally considered to own the artifacts. What he could not immediately sell he made plans to exhibit.

That exhibit opened in Piccadilly, London, in 1821, in the aptly named Egyptian Hall. The main attraction was Belzoni's fifty-foot-long model of Seti's tomb, together with full-sized recreations of two of the tomb's most impressive chambers, whose walls were covered with brightly painted bas-relief sculptures. These he reproduced in plaster, using wax molds he had made of the tomb's walls. He painted them himself, working from drawings he made of the originals. The galleries were also filled with statues, mummies, idols, scarabs, papyri, and vases, together with a host of miscellaneous objects he collected during his excavations. Taken as a whole, the exhibit sought to give visitors a taste of the thrill of discovery that Belzoni had felt when he first entered the great tomb.

The spectacle attracted large crowds—and at a shilling a head it produced a handsome return. When it closed, most of the exhibit's contents were sold at auction or by private contract. The models, the replicated wall sections, and whatever artifacts remained unsold were sent to Paris for a second, inferior exhibit that was poorly attended and soon began to lose money. Belzoni left Paris to pursue other projects, leaving the exhibit in the hands of an associate, James Curtin.

At this point, Smithson enters the story. He was living in Paris, and in 1823 he appears to have visited the exhibit and persuaded Curtin to give (or sell) him a small fragment from the painted wall of the pharaoh's tomb. This ancient fragment, with its precisely known provenance, inspired his article.

The fragment came from a mural that had been painted directly on the stone wall of Seti's tomb, and it showed four colors: white, red, black, and blue. In his article, Smithson gave a quick analysis of each color. The Egyptians, he noted, were famous for the brilliance and permanence of their white paint, and he was surprised when his analysis showed that, instead of the lead oxide or gypsum pigments that were used to make white paint in his time, the Egyptians had used the mineral aragonite. It made sense, because perfectly white aragonite deposits can often be found, and the fact that Seti's sarcophagus had been carved from a single piece of aragonite proved that plenty of it had been available. He wondered if the Egyptians had simply ground up some of the chips left over from making the

Apotheosis of a Pharaoh, one of the four-color bas-relief wall paintings in the tomb of Seti I. This colored lithograph appeared in Belzoni's book *Forty-Four Plates* (1820) and was made from his original watercolor drawing. Belzoni made wax impressions of the tomb's walls and then used them as molds to reproduce the murals in his exhibit. Smithson's fragment may have been one of the pieces accidentally pulled away from the tomb's wall as the wax was being removed. Courtesy of the New York Public Library.

sarcophagus and used that as their pigment. "Was this the material of their white paint?" he wrote.

Analyzing the red pigment posed a challenge; the layer of paint was so thin that Smithson couldn't scrape any off without taking some of the underlying rock. Instead, he identified it in place with a single blowpipe test. Directing the flame at the fragment's red paint and then quickly moving it away, produced a temporary but very specific color change. "By heating," he noted, "it became black, and returned on cooling to its original hue." This, he declared, identified the pigment as "red oxide of iron." Smithson's analysis appears to have started with an observation. Based on the pigment's rusty-red color, he must have suspected that it was iron oxide, a material readily available to the Egyptians. To verify that assumption he turned to a blowpipe test he had described twenty years earlier in his calamine article. "This quality, of temporarily changing their colour by heat," he had written, "is common to most, if not all, metallic oxides; the white growing yellow, the yellow red, the red black." In other words, if the red pigment turned black when gently heated by the flame, and then turned back to red when it cooled, it confirmed that it was a metallic oxide. This test, combined with the pigment's distinctive color, confirmed his suspicion. The red pigment was iron oxide.

The black layer of paint was thick enough to remove, so he didn't have to analyze it in place. It seems that it was not completely black but instead some shade of gray, as he reported that when viewed with a strong lens the paint consisted of a mixture of black and white particles. The white particles looked like the same ground alabaster he had found in the white paint, which was confirmed when washing the paint with an acid removed them completely. This was because alabaster is chemically what Smithson called "carbonate of lime" ($CaCO_3$), and when put in contact with a liquid acid (such as hydrochloric acid, HCl) it reacts to produce a gas (CO_2), water (H_2O), and calcium chloride ($CaCl_2$), which is soluble and would have been washed away. This left just the black particles, and Smithson reported that under a lens they had exactly the texture of "pounded charcoal." He confirmed this identification by directing his blowpipe at the sample and reporting that, as expected, "in the fire it burned entirely away." The use of charcoal as a pigment puzzled Smithson. In his time "lamp-black," made from burned lamp oil, and "boneblack," from charred animal bones, were the main black pigments. Did the Egyptians' use of charcoal signify an "unacquaintance" with these materials?

That left the blue to analyze, and Smithson now returned to a topic he had first studied twenty years earlier, when his half-brother brought him the small painted figure of Isis. "Blue is what most deserves attention," he wrote, a reference to the fact that the Egyptian method of making this blue pigment had been lost for almost a thousand years. Finding a suitable blue pigment for paint had been a problem for all ancient peoples. Organic materials were not permanent, and blue minerals were uncommon and hard to find. The best pigment came from the intensely blue semiprecious mineral lapis lazuli, but the mines that produced it were mostly found in present-day Afghanistan, and it was prohibitively expensive. At some point in their early history, the Egyptians found a solution, and they used it extensively to paint public buildings and homes, monumental works of art, and everyday decorative items. The fact that the pigment Smithson tested had lasted so long—he estimated that it was 3,500 years old, very close to the modern estimate of 3,100 years old—testified to its permanence.

It is now known that what the Egyptians made was a form of glass, similar to the silica-lime-soda glass made today, but from natural ingredients. No Egyptian recipe has ever been found, but it appears to have been made from sand (silica), limestone (lime), and the mineral "natron" (soda), and colored by the addition of some form of copper. The Egyptians ground and mixed these materials together, along with a little water, and rolled them with their hands into balls. When the balls were dry, they put them into covered, earthenware jars, which were then strongly heated for several hours in an oven at 850 to 1100°C. The materials would fuse together into a "frit," a glassy, partially granulated mass, which was then broken apart, sorted by color, and ground into pigment. The process required considerable skill to achieve reliable results, as the color of the glass depended on several variables. For example, having more copper than lime in the mix would produce a blue color, while less copper than lime would yield a green. Higher temperatures also tended to produce greens. Even how the glass was ground was important in determining the shade of blue. Fine particles produced a pale blue color; the color grew darker as the particle size increased.

Over time, the use of Egyptian blue, as it came to be called, spread to Mesopotamia and Greece, and finally to Rome. In the first-century BCE, the writer Vitruvius reported that the Romans were manufacturing it at Pozzuoli, near Naples, and he provided the earliest surviving account of how it was made. Unfortunately, he got it wrong. The process Vitruvius described strongly resembled that of the Egyptians, but he listed only three ingredients: sand, natron, and copper filings. He failed to mention the use of lime (Ca), and while it is possible that the particular sand the Romans used already contained enough lime to work, his recipe would have proved useless to anyone trying to follow it at another location. The manufacture of Egyptian blue faded out with the decline of the Roman Empire, around 500 CE.

Artists in the Middle Ages struggled to find a suitable replacement for Egyptian blue. By Smithson's time that need was largely met by "smalt" (ground cobalt glass) and "Prussian blue" (a synthetic pigment, $C_{18}Fe_7N_{18}$), but there was still interest in the Egyptian pigment. That interest was amplified by Napoleon's Egyptian excursion, which brought blue-colored artifacts back to Europe, and by the later discovery of caches of blue pigment at Roman archaeological sites. In the first two decades of the nineteenth century these materials were studied and analyzed by some of the best chemists of the time, including Klaproth, Chaptal, and Davy—all of whom Smithson knew.

By the time Smithson acquired his fragment, the use of copper in the blue Roman pigments was well established. It was also known that they had also used blue glass pigments made with cobalt, but the question of whether the ancient Egyptians used cobalt, copper, or both was still undetermined. As late as 1821, the authoritative French publication *Description de l'Égypte* declared that the question had yet to be answered. Conflicting evidence and the problem of dating the Egyptian artifacts on which pigments were found made finding a definitive answer problematic. So, when Smithson published his analysis of an Egyptian sample with a clear provenance and precisely known date of origin, it was of more than just passing interest.

Smithson's analysis of the pigment, made with a single blowpipe test, showed that "its tinging matter was not cobalt, but copper." Although its color was, to him, "a little paler" than cobalt glass, the experts he showed it to did not notice the difference. "What its advantages would be above that from cobalt," he wrote, "it is for artists to decide," but the point was moot since the process for making it remained lost. Although using copper seemed preferable to the more expensive cobalt, he reported, "I am informed that a fine blue glass cannot at present be obtained by means of copper." At another time in his life he might have performed experiments to find the recipe for Egyptian blue or offered some thoughts on how it was made, but there was no mention of this now.

Although his analysis contained no great discoveries, Smithson's article was a solid contribution to an established scientific topic, and summaries of it appeared in scientific publications like the *Bulletin des Sciences Technologiques* (1824) and the *Polytechnisches Journal* (1824). It was also of enough general interest to be reprinted in the *Gentleman's Magazine* (1825), and in the "Pigment" entry in the *Encyclopaedia Londinensis* (1825). Many years later, in its *Annual Record* (1875), the Smithsonian Institution reported that a French chemist had finally succeeded in producing Egyptian blue in his laboratory; however, there was no mention that the institution's founder had also written on this topic.

The completion of his article on Egyptian colors effectively marked the end of Smithson's research career. He would write and publish three more articles, but none would require new experiments, and none presented new scientific findings. Although he continued to read and think, that part of his life was over.

———————+———————

20.

KIRKDALE CAVE AND PENN'S THEORY

IN 1824, James Smithson wrote an article that stands apart from anything else he published. At ten pages, it was by far the longest work he had produced since leaving England, but it contained no new scientific discoveries, and it did not present any chemical analysis. Instead, the article contained Smithson's thoughts on a geological theory that had recently been published by an English biblical scholar named Granville Penn. Curiously, Smithson notes that he had not actually read any of Penn's books, but was responding to a review of one of them that had recently appeared in the Royal Institution's *Quarterly Journal*.

For more than a decade, ever since he moved to the continent, Smithson had avoided writing about geology. His 1813 article on geology likely led to his estrangement from the Royal Society, and at the beginning of this article he admits that "I have hesitated long about communicating the present observations." But ultimately, he "yielded to a sense of the importance of the subject" and made his thoughts known. What Smithson wrote was a surprising document that reveals much about him and the time in which he lived.

The story starts in the summer of 1821, when workmen from a quarry in Yorkshire discovered the narrow entrance to a long-abandoned cave. Inside they found the floor covered with a thick layer of dirt and animal bones, some of which they collected and added to the material being used to repair a nearby road. The workmen assumed the remains to be cattle bones, but a traveling naturalist noticed them and realized that they were something else entirely. What the workmen found turned out to be the fossilized remains of animals not previously known to have lived in England, including hyenas, elephants, and rhinoceros. The bones of twenty-two kinds of animals would eventually be identified among the remains in the cave. An important find, it soon came to the attention of William Buckland, the reader in geology at Oxford, and after a colleague showed him some of the cave's specimens, he was keen to see it for himself. He was not able to visit the site until midwinter, and by that time most of the cave's contents had been removed by collectors and curious visitors. Fortunately, Buckland was eventually able to examine several of those collections, and enough remained in the remote parts of the cave to allow him to make an analysis.

He found that the cave's bones had been embedded in a layer of silt, which had been underlain and overlain by thin layers of stalagmite, the dissolved mineral that slowly but constantly dripped from the cave's ceiling. From this he concluded that the bones had already been present in the cave when they were covered by silt, swept in by a great flood that had also sealed the cave's entrance with rocks and gravel. To Buckland, this meant that the bones themselves were "antediluvial"—having been in the cave prior to the flood—and the thin rocky layer of stalagmite on top of them was "postdiluvial"—having been deposited after the flood receded. Based on the rate at which the stalagmite accumulated, he estimated that the flood occurred a few thousand years earlier.

Buckland had also explored caves in Europe and believed that the flood that affected Kirkdale had been part of a massive, worldwide "deluge" that affected them as well. The idea that a catastrophic, global flood divided the Earth's history into two separate (and short) epochs was a narrative familiar to every geologist of this period, and the nature of that flood (or floods) was one of the core geological questions of Smithson's time. Buckland had previously argued that the great flood of Noah, described in the Bible, was a real event that could be confirmed by the study of the Earth, and the Kirkdale site seemed to many to be just that kind of confirmation. Buckland was careful not to make that claim in his report to the Royal Society, but his repeated use of the terms deluge, antediluvian, and postdiluvial—all of which would have been understood as references to that narrative—would have left no doubt among his readers that this was what he meant.

The question of whether geology supported biblical history was still being debated in Smithson's time, although the topic rarely appeared in scientific journals. It was an old argument, and if Buckland's analysis of the cave had only

The entrance to Kirkdale Cave, from the paper William Buckland submitted to the Royal Society, titled "Account of an Assemblage of Fossil Teeth and Bones of Elephant, Rhinoceros, Hippopotamus, Bear, Tiger, and Hyaena, and Sixteen Other Animals; Discovered in a Cave at Kirkdale, Yorkshire, in the Year 1821." One of the puzzles of the site was explaining how the bodies of so many large animals could have entered the cave through this narrow opening. Courtesy of the Smithsonian Libraries.

consisted of this finding it probably would not have attracted much attention. But Buckland's study of the cave's bones also told another story, and this one was of great interest. Most of the bones he found were from hyenas, but the cave also contained bones from a remarkable mixture of other animals, including elephants, rhinoceroses, hippopotamuses, horses, oxen, deer, tigers, and bears. The condition of the bones, few of which were completely intact, as well as the fact that they had somehow come into the cave through its small entrance led him to conclude that the cave had long been used as a hyena den. Buckland speculated that most of the animals died nearby and the hyenas then dismembered the carcasses and dragged them into the cave to be consumed. As evidence of this he noted the presence in the cave of what he called "album graecum," which he identified as hyena feces, and he reported that an analysis by the noted chemist (and Smithson's friend) William Wollaston revealed this to be consistent with a diet high in bone material. Buckland used points like this, researched in intricate detail, to reconstruct an ancient world where unfamiliar creatures roamed the land now known as England. No detail was without meaning, and the discovery of innumerable water-rat teeth in the cave led him to famously conclude that a lake, now long vanished, had once existed close to the cave, and that the hyenas sometimes hunted the foul-tasting rats when no other food was available.

Buckland's detailed reconstruction of the time of the hyenas was unprecedented and of great interest. His article was so long that it took three successive meetings to read it to the Royal Society, and the attendance at each of those meetings was said to be "enormous." The article was quickly printed in the society's journal, and later that same year Buckland was awarded the society's prestigious Copley medal "for his Paper on the Fossil Teeth and Bones discovered in a Cave at Kirkdale." At the time the award was announced, it had been less than a year since he first visited the site. The article's importance was also recognized in Europe, where summaries and translations promptly appeared in leading scientific journals. Smithson, who was living in Paris when the article came out, reported that he too had read it.

Not everyone agreed with Buckland's conclusions, especially in England. The so-called biblical literalists, while happy to have the Kirkdale site confirm the reality of the Deluge, were troubled by Buckland's description of both the flood itself and the world that preceded it. For them, the suggestion that England had once been populated by tropical animals like elephants, rhinoceroses, and hyenas seemed improbable, and Buckland's reconstruction of an antediluvian world in which hyaenas had populated a cave for hundreds of years seemed to suggest a much older Earth than their understanding of the Bible could support. There was also the problem of the Kirkdale cave being part of a limestone formation. Many literalists believed that the limestone deposits so commonly found around the globe were leftovers of the great flood described in Genesis. They argued that this material had been churned up from the ocean bottom as the waters rose and then deposited on the land as the waters receded. These thick deposits had been soft at first, which allowed shells and bones to be embedded in them before they dried into the fossil-filled stone seen today. But Buckland's assertion that hyenas had been living in caves in the limestone prior to the great flood was incompatible with this interpretation. If the limestone formation existed prior to the great flood, it implied that there had been other floods, which challenged the Bible's description of a single great Deluge.

One of the first to offer an alternative to Buckland's interpretation was Granville Penn. A well-regarded English scholar mostly known for biblical commentary and translations of classical works, Penn's first book on geology had just been published. Titled *A Comparative Estimate of the Mineral and Mosaical Geologies* (1822), the book was notable for its assertion that practitioners of scientific geology ("Mineral Geology," as he called it) were "cavillers and sophists, whose policy it is to challenge a perpetual warfare" against the kind of geology that took the Bible as its starting point (which he called "Mosaic Geology"). Penn argued that attempting to address the arguments of the mineral geologists, was playing into their "stratagem" and could only lead to "infidelity." The kind of geology Penn advocated saw the biblical narrative as sacred truth and rejected any suggestion that challenged a literal interpretation of it, no matter how strong the evidence.

Penn's book and Buckland's article were published at almost the same time, which meant that Penn had already finished writing when the Kirkdale discovery was announced. This explains why his book had no mention of the cave, even though Buckland's thesis seemed incompatible with Penn's literal interpretation of the Bible. It was a serious shortcoming that at least one reviewer called on him to address.

Penn rushed to respond and the following year, in 1823, he published a 190-page *Supplement* to his book. In it, he proposed a radically different explanation for the Kirkdale site. The bones, he said, were from animals drowned in the great flood. In the nearly twelve months that the flood lasted, their bodies had floated from the tropical zones where they had perished, north toward England. The poor condition of the bones was the result of the inevitable jostling and collisions the bodies encountered along the way. Deposited as the flood waters receded, they sank deep into the thick mud that the waters left behind, and the mud then slowly dried into limestone. Gases emitted by the animal's putrefying bodies had created a bubble around them, which hardened into a cave as the limestone dried. Penn's ingenious theory seemed plausible to many, including the reviewer for the *Quarterly Journal*, who wrote the glowing evaluation that Smithson felt compelled to oppose.

It is important to remember that Smithson was responding to the review of Penn's book rather than to the book itself. There were arguments that Penn used against Buckland's hyena theory that Smithson did not address because the reviewer had failed to mention them. What he did address were the "mineral" parts of Penn's theory: the questions of where the limestone came from, how the cave formed, and the origin of the bones.

His introduction first began with the observation, "in a book held by a large portion of mankind to have been written from divine inspiration, an universal deluge is recorded." This was a clear reference to the Bible, and Smithson observed that it was natural for those who believed in the literal truth of the Deluge to use it to explain geological phenomena—but that these explanations had failed "to obtain the general assent of the learned." And now, he wrote, "Mr. Penn has endeavoured to reconcile it [the biblical narrative] with the facts of the Kirkdale Cave."

To a modern reader, used to thinking of science and religion as separate endeavors, it might seem that Smithson is talking about any attempt to link geological facts with the history of the Earth told in the Bible. However, the idea that the Bible could be confirmed by the "book of nature" was a widely held view in Smithson's time, and few in the scientific community would have dismissed the possibility. What Smithson was really talking about was a branch of geology that took as its starting point the absolute truth of the biblical narrative. The biblical literalists believed that the Earth had been created in six days and that the Deluge happened exactly as described in the sacred text. They accepted that the flood covered the entire Earth, that all the people and animals perished—except for the occupants of Noah's ark—and that the evidence of the Deluge could be seen everywhere, if one knew how to look.

The driving force behind literalist geology in England was the distinguished natural philosopher Jean André de Luc. He had not been the first to propose this, but he was the first to present a scientifically plausible argument for it. During the French Revolution and England's subsequent war with France, de Luc persistently and effectively argued for his geological ideas and against the ideas of others. But by the time of his death in 1817, new discoveries had made his theory outdated and the ranks of his supporters had shrunk dramatically. Smithson did not identify de Luc by name, but de Luc's theory is clearly what he had in mind when he wrote that the success of the literalists "was not such as to obtain the general assent of the learned." If this characterization of de Luc seems unkind, it should be noted that Smithson had few reasons to be otherwise. De Luc had made the strongest and most personal attacks against Smithson's friend, the geologist James Hutton, and de Luc had done everything in his power to discredit Hutton's geological theory, a theory that Smithson supported. De Luc had also attacked Smithson for his "Saline Substance" article, an attack that may have contributed to Smithson's estrangement from the Royal Society.

Although the review of Penn's *Supplement* that Smithson read failed to mention it, Smithson correctly identified Penn as a follower of de Luc. As one London journal noted, "it will be seen that Mr. Penn's theory is not only the same in substance with M. de Luc, but in many instances his sentences are paraphrases upon that author." Smithson found much to oppose in the argument that Penn made to reconcile literalist geology with the Kirkdale findings. But this was also an opportunity for Smithson to engage with a supporter of his old adversary, which may explain the unusually aggressive tone of his article.

The first thing Smithson addressed was Penn's theory about limestone. One of the questions that any geological theory of that time had to answer was how limestone deposits far from the sea came to have seashells and the remains of other marine animals embedded in them. Penn's theory, as the review repeatedly noted, answered this question by proposing that the limestone materials had originally been "soft and plastic, when they rested in enormous masses, on the bed of the primitive ocean." But the flood waters stirred up the material and carried it in suspension, eventually depositing a thick layer on the land as the water receded. The shells and bones that the waters carried were also left behind, and these sank deep into the soft mud. Exposed at last to the air and the heat of the sun, the mud hardened into limestone.

Smithson agreed that the material that formed the limestone had originally collected on the sea bottom and once been soft. But he was skeptical about the flood waters stirring it up and holding it in suspension. "What navigator has told of the storm in which the sea became thick with its own sediments," he asked, and what evidence was there of a storm associated with the Deluge? "No hurricanes, no tempestuous winds, no swollen billows are recorded," he wrote. Besides, the ark "a vessel, bulky beyond all the efforts of imagination to figure, so laden, so manned, could not have lived in any agitated sea."

Even if the sediment had somehow been lifted up and deposited on the land, he argued, it probably would not have remained. The action of the waves, combined with the tides, would have carried it from the higher places, and the final rush of the receding flood waters would have carried away whatever was left. Even if the waters of the deluge did leave it behind, "They must have done so over the entire earth. It must have been a universal stratum." Which meant that "if over the earth were spread such a layer of mire, Noah and the animals could not have landed on it."

These commonsense arguments put Penn's theory in doubt, but Smithson's most telling criticism came when he turned to the question of how limestone forms. Penn had repeatedly described the "mud" left behind by the flood as "drying" to form limestone. But as a mineralogist, Smithson knew that making limestone takes more than that. Limestone is made primarily from particles of "calcareous matter" (calcium carbonate, $CaCO_3$), the same material as seashells, and if a paste of that material simply dried, nothing would hold it together. It would soon crumble and fall apart. Smithson pointed out what Penn apparently had not known, that limestone consists of particles and that they need something to hold them together. "Each particle of powder," he wrote, "is a diminutive pebble, and an intervening cement is required to connect it with the neighbouring ones." Surprisingly, experiments had shown that this "cement" was also calcium carbonate, the same material as the particles themselves. Smithson described a process in which tiny underwater "fragments of shells" were coated with this material, which precipitated onto them out of seawater. In some cases, this adhesive material could actually be seen. Smithson wrote that "in limestones consisting of considerable-sized fragments of shells, the sparry cement which connects them is perfectly evident. It is this cement which appears as regular crystals where cavities occur in the mass too large to have been filled by it." The distinctive kind of large-particle limestone that Smithson was describing is called "oolitic" limestone, and it was thought to only form under water. Smithson pointed out, "The Kirkdale rock being composed of oolites must have had this origin."

This put Penn's theory in doubt, but Smithson wasn't finished. In case Penn simply incorporated this new information by changing his theory to say that the limestone had formed underwater before the flood waters receded, Smithson described the chemistry involved. The calcareous matter that makes limestone, he wrote, is dissolved by water that contains an acid, and in sea water that acid is "carbonic acid" (H_2CO_3), which comes from volcanoes. Once dissolved, the calcareous matter will remain in solution as long as the acid is present, but under some circumstances the acid will begin to escape into the atmosphere in the form of carbon dioxide gas (CO_2). When this happens, a corresponding amount of calcareous matter will come out of solution and crystalize on the particles, binding them together to make limestone. Under normal conditions the concentrations of these chemicals in sea water is quite low, and it takes "centuries of centuries" (as Smithson put it) to accumulate enough limestone to form a bed like the one at Kirkdale. In case Penn wanted to argue that the special conditions during the year-long flood had increased those concentrations and accelerated the process down to a few months, Smithson pointed out that this created a new problem. Speeding up the rate of limestone production that much, he said, would also greatly increase the amount of suffocating carbon dioxide gas released into the air. "It is also utterly impossible," he wrote, "to believe that the beings in the ark, already not a little inconvenienced for respiration, could withstand the suffocating effluvium."

Smithson also questioned Penn's theory about how the Kirkdale cave had formed. Penn proposed that after becoming completely embedded in the mud, the bodies of the animals began to decompose, and that gases from their decomposition formed the cave in which they were found. Penn's reviewer gave few details beyond these, but Smithson was happy to list what still needed to be explained.

He first asked about the exact mechanism of how the cave formed. "If the limestone pulp was too thin, the gas would pass through it and escape," and, "if too thick, the elastic force of the gas would be insufficient to repel it." Smithson asked why the gas had formed one large bubble rather than a series of smaller bubbles around each body. And why hadn't some of the mud found its way into the mouths or exposed cavities of the animals' bodies? After their tissue rotted away, there should have been pieces of limestone left behind, "bearing the impression of the parts with which they had been in contact; as at Pompeii." If the gas pushed the mud away from the bodies, as Penn suggested, the mud "would carry with it some adhering fragments of them—bones, teeth, hair, feathers . . . which would now be fixed to the sides and roofs of the caves." Nothing like this had been found.

Smithson also asked about the nature of the gas itself, because bodies that had been floating in water for twelve months would already be partly decomposed. Even if they were whole, being embedded in the mud would have meant that they were sealed off from the atmosphere. How much gas would they have produced under those conditions? Smithson revealed he had once investigated this question: "From some experiments made a great number of years ago, on the decay of animal muscle confined over mercury [sealed off from the air], I am inclined to believe, that in no case, when secluded from oxygen, is any great volume of gas evolved by it." This hard, experimental evidence challenged one of the main assumptions of Penn's theory, and Smithson still wasn't finished. He noted that rather than decay, at least some of the flesh that was sealed off from the air and under pressure was likely to have been converted to "adipocire"—a fatty substance that animal tissue forms in anerobic conditions—but nothing like this had been found in the Kirkdale cave.

Smithson now turned his attention to Penn's theory about where the bones came from. Buckland's hyena-den theory implied that all the animals originally lived in England, but Penn rejected this. Elephants, rhinoceroses, and hyenas were tropical animals, he argued, and they could never have survived the cold English climate. But Smithson pointed out that there was evidence to the contrary. Recently, the frozen remains of an elephant (what we now call a mammoth) had been discovered in Siberia, and, as Smithson said, it was known "to have been a very hairy animal." It "may be supposed to have been a northern one," he wrote, "and if there were formerly northern elephants, there may have been northern hyaenas and northern tigers." He also questioned the likelihood of floating animal bodies acting as Penn had described. What mechanism caused them to come together as they floated north? "If they sunk below the surface, they would sink to different levels; borne on the surface, they might assemble together, but no adherence would take place between them, and upon the slightest impulse they would part again." He asked about the violent collisions that had broken the animals' bones, for any force that could break bones would also have pulverized their flesh, tearing the bodies apart. And why were there so many bones and teeth from water-rats? If they floated in the water, smaller animals would decompose first, and their bones and teeth would soon sink to the bottom. Why weren't there any remains from "reptiles, insects, trees, even fish, for all of them must have perished."

A particularly telling question was how the delicate balls of "album graecum" could have remained intact after all that time in the water. That they could have survived under these conditions, he wrote, "Will not, under any circumstances, be easy to admit." Finally, Smithson asked why—if all the world's limestone had been formed from this single event—had no one ever found the remains of land animals embedded in it? Why had no one ever found "even the slenderest bone of a water-rat bedded in the solid stone? What limestone stratum has astonished the learned, by presenting them, in its substance, with an antediluvian hyaena's bristles?"

Taken as a whole, Smithson's remarks challenged Penn's theory in a way that would have been almost impossible for him to answer. But these were scientific arguments, based on physical evidence and scientific principles, and Penn had already stated that such objections should be dismissed by the faithful as mere distractions. And so, in a section of his article titled "Of the Deluge," Smithson set science aside and reasoned directly from the sacred text. Ultimately, he said, "two great facts" proved that the Kirkdale cave and its contents had not been products of the biblical Deluge. "One," he said, "is the total absence in the fossil world of all human remains of every vestige of man himself and of his arts." Here Smithson was referring to the human victims of the great flood, which the Bible said had killed everyone on Earth except the occupants of the Ark. Smithson wrote: "Human bodies by millions must then have covered the waters; they must have formed a material part, if not the principal one, of every group [of remains], and human bones be now consequently met with everywhere blended with those of animals. Objects of human industry and skill must likewise continually occur among the bones. Of the miserable victims of the disaster numbers would be clothed, and have on their persons articles of the most imperishable materials; and the dog would retain his collar, the horse his bit and harness, the ox his yoke." If Penn's theory was correct, all of these artifacts would have been buried in the mud, and "every limestone quarry should daily present us with some of these most precious of all antiquities." But not a single one had ever been found.

"The other great fact," Smithson wrote, "which forcibly militates against the diluvian hypothesis is, that the fossil animals are not those which existed at the time of the deluge." Smithson was referring to the Bible's description of the Ark's carrying a pair of each of the world's animals at that time. If that was the case, they should have been identical to the animals outside the Ark, which perished in the flood and became embedded in the limestone. And, since modern animals are descended directly from those on the Ark, they must be identical to them, and they should also be identical to the animals whose remains are found in the stone. But, he reported, of all the fossils that had been found up to that point, not one of those animals, not "quadruped, or bird, or fish, or shell, or insect, or plant, is now alive."

Given the complete lack of evidence for what the Bible itself seemed to predict—if Penn's theory was correct—Smithson concluded that there was only one way that the theory could be supported. That was if, after the Deluge was over, the Creator performed a miracle "which swept away all that could recall that day of death when 'the windows of heaven were opened' upon mankind." Because only such a miracle could explain the lack of evidence for Penn's theory. "To a miracle then," Smithson concluded, "must we refer what no natural means are adequate to explain."

Smithson was employing a debate technique that logicians call *reductio ad absurdum*—"reduction to absurdity." He proposed a miracle because he knew that neither the religious nor the scientific communities would accept it. Literalist geologists could hardly rely on a miracle that was not mentioned in the Bible, and one of the requirements of a scientific theory is that it needs to be based on natural laws and principles *currently in effect*. The strategy seems to have worked; Penn acknowledged and responded (in some way) to most of his other critics but he never mentioned Smithson.

Smithson published his article in the *Annals of Science*, a leading scientific journal, but most of the debate about Penn's theory actually took place in the popular press. The *British Critic*, for example, published a blistering evaluation of Penn's geological work. It began by saying, "We have always doubted the expediency of connecting the speculations of science with the truths of revealed religion; and the work now before us has fully justified all our scruples on this head." By contrast, the *Literary Gazette* praised Penn for showing that the newly discovered fossils "ably and unanswerably added to the demonstrations of the truth of the *Sacred history of a deluge*." A number of popular geology books from around this time either endorsed Penn's theory or spoke highly of it. Although Penn continued to have supporters until well into the nineteenth century, they were increasingly drawn (as one historian put it) from "among those who had a superficial knowledge of the subject."

While this characterization may sound uncharitable, it was not inaccurate. Penn's theory was never seriously considered by the scientific community. Penn was not a member of any scientific society, he had not engaged in scientific research or made any discoveries, and his approach to geology was fundamentally more literary than scientific. He was unaware of some basic scientific facts and unwilling to let his theory be judged solely on the physical evidence. As a result, there was almost no mention of his theory in the scientific literature. It was briefly discussed—and dismissed—by the *American Journal of Science*, as part of a review of

Buckland's work, and derided as amateurish in a summary of Smithson's article that appeared in the *Bulletin des Sciences*. Apart from that, the only other mention of Penn in a scientific journal were Smithson's article and the highly favorable review of Penn's *Supplement* that had appeared in the *Quarterly Journal of Science*, the journal of the Royal Institution. This was the article that had induced Smithson's comments, and the fact that it appeared in the Royal Institution's journal provides a clue to why he found it so objectionable.

One of the questions that Smithson's article inevitably raises is why he wrote it when he did. Works of literalist geology were nothing new in England. His old opponent Jean André de Luc, for example, had been publishing them for decades, and yet Smithson had never chosen to express his opposition. Penn's theory would have been particularly insufferable to a man like Smithson, but if he wanted to seriously oppose it, he might at least have read Penn's book. Smithson tells us that he hesitated for a long time before submitting his article but that, ultimately, he "yielded to a sense of the importance of the subject *in more than one respect*." One of those respects may have been related to the unique role that the Royal Institution played in the first decades of the nineteenth century.

During this period the Royal Institution was perhaps the only place in England where new ideas about geology could be presented and discussed solely on their scientific merits. Humphry Davy had lectured there on his new geologic theory and, even though very little of what he said ever appeared in print, the institution provided an important venue for him to present it. As one of the institution's founders and a longtime member of its chemistry committee, Smithson valued the Royal Institution's role in scientific discourse. It is not hard to imagine the concern he must have felt as he sat in Paris reading the glowing, uncritical review of Penn's *Supplement* in the institution's journal. He wasn't the only one who found the *Quarterly Journal's* support of Penn strange. A year earlier the *Journal* published an effusive review of Penn's first book, which even some of Penn's supporters found to be excessive. The *London Journal of Arts*, for example, which characterized Penn's book as "deserving of public attention," felt obliged to also note the "extraordinary criticism given of this work by the Editor of the *Quarterly Journal of Science*," which they characterized as "loading him with unmerited compliments." Did Smithson write his article partly to call attention to what he saw as the falling standards at the Royal Institution? The evidence is suggestive, if not definitive.

But the article reveals other things that are certain. As already noted, Smithson discloses having done experiments on the production of gases. He was not previously known to have investigated this topic, although it was of great scientific interest throughout his lifetime. Smithson also reveals a surprising familiarity with fossils. It is now known that he had fossils in his mineral collection, and he had, it turns out, corresponded with the great French naturalist Georges Cuvier. Sometimes called the father of paleontology, Cuvier read a paper to the French Institut National in 1800, appealing to the entire Republic of Letters for help with his research. About to undertake an ambitious study of mammal fossils, he asked that people send him drawings of what they had in their collections. The results of that study, published in his important *Recherches sur les ossemens fossils de quadrupeds*, reveal that Smithson had responded to his call. During the time that Smithson had been resident in Kassel, probably around 1806, he visited the newly discovered cave at Altenstein. "Altensteiner Höhle," as it was called, lies about sixty miles southeast of Kassel and, at some point in the distant past, had been home to cave bears. Cuvier reported that a bear's femur bone, dyed black by the cave's soil, was sent to him "from Mr. Smithson, an English gentleman resident at Kassel." The Altenstein cave is nearly identical to the German caves that Buckland visited almost a decade later, and Buckland actually referred to it in establishing the context of the Kirkdale cave.

One other aspect of Smithson's article deserves attention, and that is what it tells us about his religious beliefs. He makes several revealing remarks as he develops the argument against Penn, the first being in the second paragraph, where he characterizes the Bible as "a book held by a large portion of mankind to have been written from divine inspiration." In the next paragraph he characterizes attempts to find physical evidence of the events described in the Bible as "not such as to obtain the general assent of the learned." Toward the end of the article, Smithson speaks more generally of efforts "to collate the revered volume with the great book of nature, and show in their agreement one author to both." It was, he admits, a worthy goal, but "the result has not been what was anticipated."

Like so much about Smithson, what we know about his personal beliefs is limited. We know that Oxford students had to declare their adherence to the Church of England as a condition of admission, and he may have been expected to attend services while he was there. Surviving letters, written when he was in his twenties, show him to have been critical of organized religion. At that point in his life, like many of his contemporaries with similar educational and social backgrounds, Smithson seems to have been a deist: believing in a creator but valuing reason and the laws of nature over tradition and sacred texts. The only other information about his religious beliefs comes from this article, written when he was fifty-nine. His overall tone, combined with the fact that he makes several references to the "Creator" while studiously avoiding identifying the Bible as the document being discussed, indicate that his views on this topic had not changed.

In terms of its influence, Smithson's article generated little comment. Being a response to a book review and containing no new discoveries, it was not noted in other journals. Smithson was only one of Penn's many critics, some of whom were also literalist writers, and others made many of the same points. Penn would write one more book on geology: a second, enlarged edition of his *Comparative Estimate*, published in 1825, in which he responded to his critics as best he could. After that he turned to other topics and made no further statements about geology. His engagement with the subject had lasted just three years. Smithson also avoided making any further comments on geology. His rebuttal of Penn's theory, while effective, would be remembered mostly for what it revealed about Smithson himself.

21.

THE "INCREASE AND DIFFUSION OF KNOWLEDGE"

IN THE EARLY 1820s, as Smithson approached the age of sixty, his scientific articles began to change. The articles were still of scientific interest, but now he increasingly included descriptions of the tests he used and advice on how to use them. He also wrote about the specialized tools that he used, and some of his articles were confined to that topic. He seemed to be thinking about the future and passing along his knowledge.

In 1823, Smithson wrote an article intended for a broad audience. Titled "On the Discovery of Acids in Mineral Substances," its goal was to provide a simplified method of analyzing minerals and, specifically, of identifying any acids they contained. As Smithson put it, he wanted to give his readers a way "to ascertain with certainty whether an acid does or does not exist [in a mineral], and, if one is present, its species, and this with such facility that the trial may be indefinitely renewed at pleasure, and made by all." Smithson's previous articles assumed that his readers were knowledgeable chemists, but now he seemed to be trying something new. In this article, and for the first time, he assumed the role of a teacher and addressed a basic topic—the identification of acids—in a clear and systematic way. "So far as I have gone in these respects," he wrote, "I here impart."

In Smithson's time the known acids were still grouped according to the "kingdoms" of nature in which they were found: mineral, vegetable, and animal. Here

Smithson was only concerned with the mineral acids—those known today as "inorganic." The topic of acids was an important one, because chemists of this era almost exclusively divided the substances that formed minerals into two great classes: acids and alkalis (or, as they were sometimes called, acids and bases). As Smithson's friend Thomas Thomson put it, with the exception of a few metals, "minerals consist essentially of these two classes of bodies." Although minerals were identified largely on this basis, the identification of mineral acids remained problematic. Acids, as Smithson wrote, "have been repeatedly overlooked in mineral substances, and hence dubiousness still hovers over the constitution of many, although they have formed the subjects of analysis to some of the greatest modern chemists."

When Smithson wrote this article, a large chemical literature already existed, with multiple tests available to identify each of the known acids. These highly specific tests, largely derived from laboratory chemistry, must have been daunting for nonspecialists to apply to the complicated mineral substances found "in nature." This explains why most of the mineral analyses reported in this period's scientific literature were made by just a handful of accomplished specialists. Finding a simplified method of identifying the acids in minerals was an obvious need, and one that Smithson set himself to address. The methodology he laid out in his article reflected a lifetime of experience and thought, and while he undoubtedly took shortcuts in his own work, the process he described was likely his basic approach to analyzing an unknown mineral.

It began with reducing a small piece of the sample to powder, and then adding either "carbonate of soda" (baking soda, $NaHCO_3$) or "potash" (K_2CO_3). The mixture was then subjected to the flame of a blowpipe and melted into a solid mass. This served the dual purpose of burning away some of the impurities and giving the first indication about acids in the sample. If the mixture did not melt, it did not contain an acid.

The next step was based on Smithson's observation that "lead forms an insoluble compound with all the mineral acids except the nitric," a fact that he used to make a simple, clear test for the presence of acids. As the Swedish chemist Jacob Berzelius described it many years later, once the sample had been melted "according to the method proposed by Smithson," it should be "dissolved in water, of which a clear drop is then to be let fall upon glass, saturated with acetic acid, and tested by a solution of the acetate of lead. All the mineral acids, except the nitric, give precipitates." Smithson described a separate test to determine if the liquid contained nitric acid. In other words, if any solids came out of the liquid or if the liquid became cloudy in any way, there was an acid in the mineral being tested, and the next step was to identify it.

Smithson listed eleven mineral acids in his article, and he provided straight-forward tests to identify each one. Most of these were well-known tests selected from the many existing ones in the scientific literature, but three were tests that Smithson seems to have devised himself. The first was the test for sulfuric acid, described in his article "An Account of a Native Combination of Sulphate of Barium and Fluoride of Calcium" (1820), and the second was the arsenic acid test, described in "On the Detection of very Minute Quantities of Arsenic and Mercury" (1822).

Sulfuric Acid	Molybdic Acid
Muriatic Acid	Tungstic Acid
Phosphoric Acid	Nitric Acid
Boric Acid	Carbonic Acid
Arsenical Acid	Silica
Chromic Acid	

These are the eleven known mineral acids that Smithson discussed in his article, although not everyone would have agreed with his choices. For example, he included silica, which he had been the first to identify as an acid, but for many chemists this had yet to be proved. Smithson also included phosphoric acid, although many chemists considered that to be an animal acid. Many chemists would also have included fluoric acid in this list, but Smithson would later argue that while fluorine existed in many minerals, its acid did not.

However, Smithson had not previously described the third test, for "muri-atic acid" (HCl). "I have likewise discovered a test of chlorine, and consequently of muriatic acid," he wrote. "If any matter containing chlorine or muriatic acid is laid on silver in a drop of solution of yellow sulphate of iron, or of common sul-phate of copper, a spot of a black chloride of silver, whose colour is independent of light, and which has not been attended to by chemists, is produced." The fact that this test could identify not only hydrochloric acid but also chlorine compounds was noteworthy, as was the fact that it could be applied to substances other than minerals. "The chlorine in a tear, in saliva, even in milk, may be thus made evi-dent," he wrote. Also noteworthy was the fact that the test could be used on a sam-ple smaller than a single drop of solution—because one of the few personal stories we have about Smithson involves his analysis of just such a sample.

Smithson's friend Davies Gilbert once related a story that Smithson enjoyed telling: "Mr. Smithson frequently repeated an occurrence with much pleasure and exultation, as exceeding anything that could be brought into competition with it . . . Mr. Smithson declared, that happening to observe a tear gliding down a lady's cheek, he endeavoured to catch it on a crystal vessel: that one-half of the drop escaped, but having preserved the other half, he submitted it to reagents, and detected what was then called microcosmic salt, with muriate of soda; and, I think, three or four more saline substances, held in solution."

Smithson's ability to analyze such a small sample is impressive, and the test he used to identify the muriate of soda in the tear was the same one that he now recommended for identifying muriatic acid. The social occasion in which this analysis was demonstrated is not recorded, but the fact that Smithson chose—and was able—to make this analysis speaks not only to his skill but also to the central role that chemistry and scientific inquiry played in his life, including his social life.

Most of the other tests for acids that Smithson described in his article were well known and are not of particular interest here. But the three tests that he introduced were important because of their sensitivity, and so they were adopted by chemists, especially German chemists, who tended to characterize them as "Smithson's Methode"—a term sprinkled into German mineralogical literature in the decades after this article appeared. The article was also noted in French journals: a summary of it appeared in the *Bulletin Général* (1823), a long review of it in the *Bulletin des Sciences Mathématiques* (1824), and an extract from it in the *Annales des Mines* (1826).

The article's greatest impact came in English-language publications. The entire article was quickly reprinted in the *Technical Repository* (1823), the journal of the Society for the Encouragement of Arts, in London. It was also lavishly praised in *Gleanings in Science* (1829), the journal of the Calcutta-based Asiatic Society, in which the reviewer declared, "Mr. Smithson has added largely to the subject, and to us appears to have left little to desire. His paper on the detection of the several acids, inserted in the *Annals of Philosophy*, is extremely valuable to the mineralogical student, whom it enables to make in a few minutes as unexceptionable a qualitative, if not quantitative analysis as the chemist could in several hours."

Ultimately, Smithson's article found its true home in the growing category of introductory mineralogy textbooks. His goal had been to describe reliable tests that could be "made by all," and the inclusion of his approach in many of these texts shows that he succeeded. Thomas Thomson recommended Smithson's article in his *Outlines of Mineralogy, Geology, and Mineral Analysis* (1836), and Berzelius incorporated Smithson's method of detecting acids in *The Use of the Blowpipe in Chemistry and Mineralogy* (1844), a widely read work that was also translated into German. Smithson's article was so influential that in 1929 it was republished in its entirety by *The Science News-Letter* as a "Classic of Science."

Smithson's interest in sharing his knowledge, and his success in doing so, have not received much attention, but this was clearly an important topic for him, and this interest may also explain his last two articles, which he wrote during his final visit to London.

In the spring of 1825 Smithson made his last trip back to England. He was almost sixty and had spent most of the past decade in Paris. It was time to go back, see old friends, and put his affairs in order. Despite his estrangement from the Royal Society, Smithson remained a member and clearly still had friends among the society, because on May 5, shortly after arriving, he was invited to attend a meeting of the exclusive Royal Society Club. The club's members were all from the society's inner circle and always met for dinner and conversation before the Royal Society meetings. Humphry Davy, president of both the club and the society, may well have been there, as well as Smithson's old friend Davies Gilbert, who would become the society's next president. It may have been around this time that Smithson told Gilbert the story discussed earlier about analyzing a tear.

Smithson attended the dinner as the guest of Sir George Staunton, who had been active in the East India Company and was one of the founders of the Royal Asiatic Society. The RAS, as it came to be known, had been founded just one year earlier, charged with "the investigation of subjects connected with and for the encouragement of science, literature and the arts in relation to Asia." When the first issue of the RAS journal *Gleanings in Science* came out, it contained a highly complementary review of Smithson's article, "On the Discovery of Acids in Mineral Substances."

Having finally returned to London, Smithson was in no hurry to leave. He stayed for about a year and a half, not returning to Paris until the end of 1826. In addition to renewing his scientific connections, he used this time to see the sights of London and take in all that had changed since he had left—and there was a lot to see. But Smithson was also putting his affairs in order. He met with his bankers to see to his investments, and he spent a fair amount of time going through possessions and papers that he had left in storage. As he did so, an old letter caught his eye. It was from Joseph Black, the iconic chemistry professor at the University of Edinburgh, whom Smithson met as a young man. The two corresponded for many years, and in 1790, in response to a question, Black described a scale he used for weighing small objects. The scale was simple, lightweight, inexpensive, and surprisingly accurate. Smithson had made one just like it and used it for years. Seeing the letter must have made him think about how useful it would be to others—especially students and workers, for whom professionally made scales were often prohibitively expensive.

This, then, became Smithson's next article. Dated "London, May 12, 1825," and titled "A Letter from Dr. Black Describing a Very Sensible Balance," it copied Black's letter, followed by short remarks by Smithson. The balance was simple and consisted of just two parts. One was a small rectangle of brass sheet, with two of the sides bent up into a "U" shape. The other was a strip of wood, twelve inches long, three-tenths of an inch wide, and "not thicker than a shilling."

This is the drawing Smithson made to accompany the description of Black's balance. Like a similar drawing that Black sent him, this does not show several important details, particularly the wooden beam's "transverse lines," which determined where the weights were placed. Natural philosophers of this time were often more comfortable with written descriptions than with images, and this seems to have been true of Smithson. Courtesy of the Smithsonian Libraries.

This wooden strip, which Black called the "beam," was divided into twenty parts. He wrote: "These are the principal divisions, and each of them is subdivided into halves and quarters. Across the middle is fixed one of the smallest needles I could procure to serve as an axis, and it is fixed in its place by means of a little sealing wax." Surprisingly, the needle that served as the beam's axis was placed on top of the beam, but because the beam was light and would only be used to weigh very light objects, this arrangement worked. The beam assembly itself was placed on the bent-brass "fulcrum" as seen in the drawing. The sides of the fulcrum were no more than two-tenths of an inch high, which meant that the beam's range of motion was extremely limited—hardly more than a quarter-inch up and down at the ends. But that was all that was necessary.

The weights were also simple. Black described using "one globule of gold, which weighs one grain; and two or three others which weigh one-tenth of a grain each." He also used short lengths of brass wire that each weighed one-thirtieth of a grain, and he could make even lighter ones, although he had not felt the need to do so. Chemists at this time typically used two balances, one for large and one for small weights, and Black recommended his for "any little mass from one grain or a little more to the 1/1200 of a grain." Since it takes 15.43 grains to make a modern gram and almost 454 grams to make a pound, it is understandable why Smithson found this balance perfect for the microchemistry he was developing. Its light weight and small size also made it ideal for traveling.

Unlike today's familiar beam balance, in which pans are suspended at equal distances from the center of a long beam, Black's balance was designed to make use of the principle of leverage. Like any balance, if an object weighing one grain was placed on the "ten" position on one end of the beam, it would balance an object of the same weight on the "ten" position of the other end. But it would also balance a two-grain object placed at the "five" position on the opposite side of the axis. This was because of leverage, and this feature could be utilized to weigh objects much lighter than any of the scale's weights, which increased its accuracy.

Black did not give an estimate of his scale's accuracy, only saying that "this beam has served me hitherto for every purpose." One later writer who made and experimented with one of these balances reported that it was accurate to one-thousandth of a grain. He also reported using the balance to weigh objects up to nineteen grains, although this required using a "scale pan."

The article ended with a few paragraphs by Smithson about the method he used to make very small weights from brass wire, which was an improvement over the method Black described. It was published, as were all of Smithson's articles from this part of his life, in the *Annals of Philosophy*. This one generated quite a bit of interest, though apparently not in scientific circles.

The only scientific journal that noticed it was a long, favorable review that appeared in the Royal Institution's *Quarterly Journal* (1826). The other three journals that discussed Smithson's article were devoted to practical applications of scientific knowledge. *The Technical Repository* (1826), published by the London-based Society for the Encouragement of Arts, which favored the education of working-men, and The *Mechanic's Magazine* (1827) and the *Glasgow Mechanics Magazine* (1826), both of which served a mostly working-class audience.

This emphasis on the practical value of the balance—particularly to non-chemists—continued as descriptions of it multiplied in the introductory student literature. The first to recommend it was Michael Faraday of the Royal Institution, who reprinted almost the entirety of Smithson's article in his widely read *Chemical Manipulation, Being Instructions to Students in Chemistry* (1827). Faraday's recommendation of it as "a very delicate and easily constructed balance" spread across Europe as his book was translated, first into French and then into German. As late as 1842, in the book's third edition, Faraday still introduced his discussion of it by saying, "The following account of a very delicate and easily constructed balance is from a letter written by Dr. Black to Mr. Smithson, and is a valuable piece of information to those who, having occasion for, are deprived [by cost] of the use of, a good balance."

Faraday was not alone in recommending the balance. Samuel Gray, a self-styled "practical chemist," called it "a very ingenious contrivance of a miniature steel-yard instead of the expensive scales used by assayers and philosophical chemists." He promoted it in his book *The Operative Chemist* (1828), and like Faraday's, Gray's book was translated into French. Other scientists writing for popular audiences followed suit. When physicist Henry Kater and science writer Dionysius Lardner teamed up on *The Cabinet Cyclopaedia* (1830), they included a description of Black's balance.

All these recommendations of Black's balance inevitably bring up the question of whether and to what extent it was actually used. Both Black and Smithson mentioned using it in Smithson's article, and Black made references to using it in some of his other correspondence, but did Smithson's article induce others to try it? More specifically, did students, assayers, and workingmen actually build and use balances like this? Once the materials and simple tools had been gathered, it took me just a few hours to build a workable balance like the one Black described, and a few hours more to make the weights—so making one should not have been an insurmountable challenge to nineteenth-century students or workers. But the same lightweight construction that made the balance sensitive and inexpensive also made it delicate, which may account for the fact that no known examples survive.

But there is evidence that the balance was made. In 1860, in his practical manual for chemistry students, the French author Henry Violette included a description of Black's balance. He mentioned that Smithson had been the recipient of Black's letter, then added, "This scale is easy to build, and I invite the student to follow the advice of Dr. Black." But for those students not able or not inclined to make their own, he noted that a modestly priced commercial version was also available, and he provided an illustration. Although it had been slightly modified, the instrument in Violette's drawing was clearly based on Black's design. The fact that it was still being sold so long after Smithson first described it demonstrates that students were using it. Although it was barely noticed in the scientific journals, Smithson's article was a meaningful contribution to the education of both students and adult workers.

In that regard, the timing of Smithson's article deserves further consideration. The fact that he chose to publish Black's letter at this particular time seems significant, because the letter had been in his possession for thirty-five years. If he wanted to share it with the scientific community, it seems he would have already done so. Why did he choose to publish it now? Curiously, there are similar questions about the other article Smithson wrote while in London about preserving a work of art: hardly a scientific subject, and hardly a topic of much interest to the scientific community.

As the fall of 1825 approached, Smithson was still putting his affairs in order and disposing of items he no longer needed. He may have been the "Smithson" who sold two paintings at Edward Foster's auction house on Greek Street, and in this next article he reveals that he had another artwork that he wanted to take back to Paris. He described it as a "crayon portrait," an image of someone that he wanted to preserve, "but without the frame and glass, which were bulky and heavy."

Smithson was talking about a pastel portrait. Pastels were frequently called crayons in his time, a reference to the fact that they were sold in the shape of short sticks. Pastel colors were (and still are) made from a mixture of powdered pigment, a "filler" such as powdered chalk, and a binder to hold them together. Shaped into sticks and then dried, they were almost always applied directly to a piece of paper—usually blue paper, since pastel colors tend to be light—and the paper was then pasted onto a piece of canvas, which made it easy to mount in a protective frame.

In Smithson's time pastel "paintings" were the preferred medium for portraits and highly prized for their beauty and fine, lifelike appearance. Rubbing the stick directly on the paper deposited a thick, velvety layer of dry pigment that could then be blended with other colors and worked to produce skin tones that oil paints simply couldn't match. But they were remarkably delicate. Even brushing against a pastel image could smear the pigments and ruin it, which is why they were kept in picture frames behind a protective sheet of glass.

Smithson wanted to take the portrait back to Paris, but in the early nineteenth century shipping a framed pastel with a delicate glass front would have been expensive, inconvenient, and probably risky. An oil painting could be removed from the frame, carefully rolled up, and safely transported. Smithson began to wonder if there was some way to do the same with a pastel, a way to "fix" the pigments so that it could be handled and shipped. The answer he found became the title of his article: "A Method of Fixing Crayon Colours."

He was not the first to think about this. Protecting finished pastel images had always been a concern, and before the eighteenth century the size of pastels was limited partly by the size of the available protective glass. Only after glassmaking technology reached the point that affordable plate glass was produced in the necessary sizes were pastels made on a scale comparable to easel-painted oils. Although glass provided the necessary protection, it also had several disadvantages, and over the years many artists and inventors looked for ways of "fixing" these delicate images so the glass would not be needed.

One list of pastelists working before the nineteenth century identifies more than twenty different processes that had been proposed for protecting pastels.

A METHOD OF FIXING CRAYON COLOURS.

From Thomson's Annals of Philosophy, Vol. XXVI; New Series, Vol. X, 1825, page 236.

LONDON, *August* 23, 1825.

GENTLEMEN: Wishing to transport a crayon portrait to a distance for the sake of the likeness, but without the frame and glass, which were bulky and heavy, I applied to a man from whom I expected information for a method of fixing the colours. He had heard of milk being spread with a brush over them, but I really did not conceive this process of sufficient promise to be disposed to make trial of it.

I had myself read of fixing crayon colours by sprinkling solution of isinglass from a brush upon them, but to this too, I apprehended the objections of tediousness, of dirty operation, and perhaps of incomplete result.

On thinking on the subject, the first idea which presented itself to me was that of gum-water applied to the *back* of the picture; but as it was drawn on sized blue paper, pasted on canvass, there seemed little prospect of this fluid penetrating. But an oil would do so, and a drying one would accomplish my object. I applied drying oil diluted with spirit of turpentine; after a day or two when this was grown dry, I spread a coat of the mixture over the front of the picture, and my crayon drawing became an oil painting.

Smithson's entire article about "fixing" pastel paintings. Consisting of just three paragraphs, it occupied less than half a page. Unlike any of his previous articles, this one appeared in the "Miscellaneous" section of the *Annals of Philosophy*. Had it not been for Smithson's scientific reputation and his long history as a contributor, it is not clear that the journal would have accepted it. Courtesy of the Smithsonian Libraries.

These ranged from adding wax or oils to the pigments (including forms of "encaustic" painting) to applying special varnishes to the finished image, and even to a method of sealing the image between two pieces of glass. But none of these methods proved satisfactory. The question was of sufficient importance that in London, on two separate occasions, the Royal Society for the Encouragement of Arts, Manufactures and Commerce offered cash prizes to anyone who could demonstrate a workable method. In 1768 an English artist was awarded a prize "for the discovery of his method of painting in fixed crayons"—apparently a process of mixing wax with the pastel sticks prior to using them—and several years later "an ingenious foreigner" demonstrated a method of coating a finished image with a liquid that protected it from handling. Like all of the other fixing methods, this one changed the very thing that made pastels so attractive. As witnesses reported, "The picture, which before had a particularly warm, brilliant, and agreeable effect, in comparison became cold and purple, and though in one sense the attempt succeeded to the designed intention of fixing the colours, yet the binding quality of whatever fluid was made use of in the process, changed the complexion of the colours, rendering the cold tints too predominant."

This was how things stood when Smithson became interested in the question. The first person he asked recommended brushing milk on the painting. This may have had something to do with the fact that a light-colored glue can be made from milk, but its consistency would have made it difficult to apply without damaging the image, and it may not have been permanent. As Smithson wrote, "I really did not conceive this process of sufficient promise to be disposed to make trial of it."

Smithson also reported that he had read of fixing pastels with a solution of isinglass, a substance obtained from the swim bladders of certain kinds of fish. Best known in England as a clarifying agent in beer, isinglass had long been used in Russia to make an adhesive for the restoration of paintings. In the nineteenth century it was increasingly used in Europe for paper and book restoration—and as a fixative for pastel paintings. Today a material like this could be sprayed on a painting, but in Smithson's time that technology did not exist. Instead, he reported reading that it was applied "by sprinkling [the] solution of isinglass from a brush," a process he decided not to try.

What Smithson decided he needed was something that would not only consolidate the pigments but also bind them to the paper. His first thought was to accomplish this by brushing "gum-water" on the back of the portrait. This diluted adhesive, similar to the glue now used on envelopes, seemed promising, but to work, it would have had to seep through the back of the paper that the portrait was painted on, as well as the canvas that the paper was pasted to. Smithson reluctantly concluded that "there seemed little prospect of this fluid penetrating." But, he wrote, "An oil would do so, and a drying one would accomplish my object. I applied drying oil diluted with spirit of turpentine [to the back]; after a day or two when this was grown dry, I spread a coat of the mixture over the front of the picture, and my crayon drawing became an oil painting."

Smithson's method was to use a "drying oil." In his time materials of this sort were also known as "fixed oils," because they formed a hard film when exposed to the air. These oils were extracted with a mechanical press from the seeds and kernels of plants, and the expressed oil was then processed by boiling. The most common drying oils in Smithson's time were linseed and walnut oils, and he probably used one of these, thinned with turpentine to help it penetrate. The oil did not actually dry, as Smithson described it, in the sense that water or some other solvent evaporated. Rather, after the kerosene evaporated the oil reacted chemically with oxygen in the air to form a three-dimensional network of bonds holding the film together, along with anything embedded in it—which in this case was the pastel portrait. The complex process of how this reaction happens was just beginning to be explored in Smithson's time, and even in our time not all the steps of how a film forms have been identified.

But this was not intended as a scientific article. Smithson was describing a practical technique that would mostly have been of interest to people outside the scientific community. To be sure, there was some scientific interest in what Smithson described. The London chemist Charles Smith, for example, who was a partner in a Piccadilly art supply firm that advertised a method for "fixing" soft crayon and chalk drawings, probably read Smithson's article carefully. But most of those interested in Smithson's article were artists, artisans, and adults seeking education, as can be found in the list of publications that noted Smithson's article. Other than the *Annals of Philosophy*, which seemed willing to publish anything Smithson submitted, his article was not mentioned in any of the English or French scientific journals. Only the German *Polytechnisches Journal* (1825) commented on the method employed by "Hr. Jak Smithson" to preserve drawings on paper. There was interest from magazines reporting on the fine arts, and so brief descriptions of Smithson's method appeared in England in *Bell's Court and Fashionable Magazine* (1825), in America in the *Museum of Foreign Literature and Science* (1826), and in Germany in *Der bayerische Volksfreund* (1826). Much of Smithson's article was also translated and reprinted in a German drawing manual.

Smithson's article also attracted attention from publications related in some way to adult education. In London, *Mechanics' Magazine* (1825) quickly printed a summary of Smithson's method. This publication's dedication to education for workers can be noted in the long article preceding the discussion of Smithson's method, which was titled "Education of the People," and went to great lengths to show "how groundless the apprehensions of those are, who imagine, that to make mechanics [workers] more intelligent and better informed, is to make them worse servants and members of society." Smithson's article was also reprinted in the *Franklin Journal and American Mechanics' Magazine* (1826), an American journal founded with the same goal as its English namesake. The English *Register of Arts and Journal of Patent Inventions* (1828), dedicated to keeping workers informed about new technology, also printed a summary of Smithson's article, as did two books on "science recreations"—a literary form designed to make it easy for workers and students to educate themselves about practical science.

Only in the second of these books, *Recreations in Science* (1830), did anyone comment on how well Smithson's "fixing" method worked—and that it wasn't perfect. Once the "drying oil" hardened, it did indeed protect the painting from damage, but the image itself was certainly changed, and the author of *Recreations* noted that "this process will, no doubt, effect the object." At least some of the colors would be darkened, and much of the warmth of the painting would be lost. For Smithson this does not seem to have mattered. The portrait he treated was probably of someone he knew and, as he said, he wanted it "for the sake of the likeness," not as an artistic work.

Reading between the lines, what Smithson seems to have enjoyed about this endeavor was using familiar materials—the thinned drying oil—to do something that no one had previously thought to do: turn a pastel into an oil painting. And this seems to have been what made the story appealing to the worker-education journals as well. It showed what knowledge and ingenuity could do with common materials.

The remaining question about this story is: Why did Smithson publish it? There is no new science here, no new discoveries, no new tools, no new insights. The traditional view of Smithson's science has long been that he was either a dabbler, a well-intentioned amateur working on a series of obscure topics, or that he was a second-tier figure, technically skilled and always hoping to make a big discovery, but never quite reaching that goal. If any of these assessments seems fair, then this article can be seen to confirm them.

But there is another possibility. Whatever else Smithson was, he was an idealist, and he carried the ideals of his youth throughout his life. As an "Enlightened" man, he believed that science should ultimately better the human condition and perhaps he saw publishing this story as a small way to do that. But what was going on in London, in 1825, that would have led Smithson to think about this? And who would have been the beneficiaries?

One of the big stories in England in the 1820s was the push to educate the working class. Most of the impetus came from the workers themselves who were hungry for social and economic opportunity, and for the most part it took the form of establishing worker-education societies called Mechanics' Institutes. For a modest annual fee, these locally run organizations provided workers with libraries, reading rooms, encouragement, and instruction in a wide range of useful topics—including chemistry—that offered the promise of advancement. The movement originated in Glasgow and quickly spread to London, which, by 1825, had become its de facto center.

In 1825, as Smithson refamiliarized himself with London, it would have been hard for him not to be aware of this movement. The London Mechanics' Institute chapter had been founded a year earlier, when two thousand workers packed into a meeting at the Crown and Anchor Tavern (the same place where the Royal Society Club met) to elect their first president, George Birkbeck. Smithson may have known Birkbeck—or known of him—since both men had ties to the Royal Institution, and Birkbeck had briefly been the editor of the London journal *The Chemist*, where he reviewed one of Smithson's articles. Birkbeck was active and well respected, and his name was frequently in the newspapers, as was the Mechanics' Institute. The institute's new headquarters was nearing completion, and with a library, classrooms, teaching laboratories, and a lecture hall that could hold a thousand people, it was close in size to the Royal Institution. As such, it promised to be an important new social force in the city—a prospect that not everyone viewed with approval.

English conservatives still feared the kind of working-class empowerment that had, in their view, contributed to the French Revolution only a few decades earlier, and there was a lively discussion in the press as to how worker education should be funded, organized, and controlled. In 1825, two alternatives to the Mechanics' Institute (with different management and funding) were also established in London: the City of London Literary and Scientific Institution and the Western Literary and Scientific Institution. There is no indication that Smithson was involved with any of these organizations, but he was inquisitive and read the newspapers. This was an important topic at the time, and he likely had a general sense of what was going on.

The Mechanics' Institute movement is also important because of the special terminology that became associated with worker education during this period. Of particular interest is the insistence of men like Birkbeck on talking about "the diffusion of knowledge." This phrase had for many decades been used in a general sense to refer to public education and the generalized benefits that would accrue from it. But in 1820s England, this phrase took on a more political significance. In the hands of the Mechanics' Institute movement, the lack of access to knowledge began to assume the status of a class barrier, and the successful "diffusion of knowledge" as the removal of that barrier. In the speeches made to the large crowds that assembled for the opening of the London headquarters of the Mechanics' Institute in July 1825, there were repeated calls for the "diffusion of knowledge," and the dedication carved into the institute's foundation stone began: "Now have we founded an edifice for the diffusion and advancement of human knowledge." Similarly, in the dedication of the two other worker organizations founded in London in 1825, it was made clear that the object of both was the "diffusion of useful knowledge amongst persons engaged in commercial and professional pursuits." One of Birkbeck's friends even founded a worker-education organization named the Society for the Diffusion of Useful Knowledge, which promoted the self-education of the English working class.

An awareness of these organizations and their common language is important not only to the context of Smithson's article. It also provides context for the language he used in another document he created while in London—his will. Smithson's will has always been something of a puzzle because most of it consisted of ordinary instructions for the disposal of his estate. Only at the end, in the unlikely event that his heir died without producing children, did he direct that his fortune go "to the United States of America, to found at Washington, under the name of the Smithsonian Institution, an establishment for the increase and diffusion of knowledge among men."

This single sentence was the full extent of Smithson's instructions, and the precise meaning of "increase and diffusion of knowledge" was the subject of much discussion in the 1840s when the Smithsonian was being formed. In the succeeding years a broad interpretation of "diffusion" has allowed the Smithsonian great flexibility in the topics it studies and in its vast outreach—all things that it seems certain Smithson would have approved. Nevertheless, as can be seen from this brief overview, it seems likely that these three articles were written to support the Mechanics' Institute movement, and there is suggestive evidence that Smithson's intention in writing his will was that the funds be used specifically for some form of worker education.

———————————|———————————

EPILOGUE: WHO WAS
JAMES SMITHSON?

SMITHSON WROTE AND SUBMITTED his will in the fall of 1826, and by 1827 he was back in Paris. He would never see England again, but life in Paris had its advantages. One was the vibrancy of its scientific community. A side-effect of Napoleon's long reign had been to reenergize French science, particularly French chemistry, and many of the best chemists in the world now lived and worked in Paris. Smithson knew and socialized with many of them. Another advantage was that the state of the French postwar economy meant that Smithson's money went further than it would have in England. In France he could live a lifestyle that would have been unavailable to him in England. But while his wealth allowed him to live in comfort, that life was not easy, for he was in constant pain.

A forensic analysis of his remains, undertaken by the Smithsonian in the 1970s, explains why. It estimated he had been about 5'6" and of slight build. He lived a relatively healthy and active life. But in his later years he suffered from arthritis, and there were "slight variants in his spine" that likely led to lower back pain and noticeably affected his gait and stance. There was also evidence of arthritis in his neck, which the report suggested was the result of sitting for long hours, bent over a desk, studying. Dental problems were common in Smithson's time, and by the time he died, he had lost seventeen teeth, while five of his remaining teeth were abscessed. Smithson also appears to have smoked tobacco using a short clay pipe, as was common. Pipe-stem wear (called a pipe facet) can be seen on his teeth on the left side.

Smithson's medical problems made it difficult for him to sleep, and on many evenings his solution was gambling, which acted like a palliative and allowed him to get a few hours of rest. It was a lifestyle that required him to be largely nocturnal, and in a letter written to his friend Berzelius when Smithson was fifty-three, he described a typical day: "I keep such very late hours & generally rise so late that I have indeed no day left." Both his physical decline and his response to it were noticed by his colleagues. François Arago, one of France's most distinguished scientific figures, described meeting him around this time: "In Paris I made the acquaintance of a distinguished foreigner, of great wealth, but in wretched health, whose life, save a few hours given to repose, was regularly divided between the most interesting scientific researches and gambling. It was a source of great regret to me that this learned experimentalist should devote the half of so valuable a life to a course so little in harmony with . . . intellect." Smithson lived like this for years, balancing pain with distraction and struggling to remain scientifically active. As his health declined this became increasingly difficult, and he seems to have stopped scientific research completely by the time he returned from London.

Smithson stayed in Paris until the summer of 1828. He had spent much of his life traveling and now set off on one last trip. He traveled in style, in his own carriage, with his servant and many trunks filled with his possessions. He did not take his scientific instruments or his mineral collection.

He went to Genoa, presumably to rejuvenate his health. Still, he seems to have sensed that the end was near, for among his papers he carried a copy of his will. Genoa was a comfortable spot for the wealthy, and Smithson spent several months there, tended by his servant. He passed away on June 27, 1829, at the age of sixty-four. No details are known about his passing or the cause of death. The British vice consul was called in to make the necessary arrangements, including having Smithson buried in a Protestant cemetery on a site overlooking the Ligurian Sea.

It took time for news of Smithson's death to reach England, and it was not until November 30, 1829, that his passing was announced to the Royal Society. As part of his anniversary address, the society's president, Davies Gilbert, read eulogies for deceased members, including Smithson. They had met while students at Oxford, Gilbert reported, and had been lifelong friends. He noted Smithson's friendship with Henry Cavendish and that Smithson had also been a friend of William Wollaston, "and at the same time his rival in the manipulation and analysis of small quantities." This is also where he told the story of Smithson analyzing a woman's tear, and he closed with this observation: "For many years past Mr. Smithson has resided abroad, principally, I believe, on account of his health; but he carried with him the esteem and regard of various private friends, and of a still larger number of persons who appreciated and admired his acquirements."

By "acquirements" Gilbert presumably refers to Smithson's skill as a chemist, and his affection and respect for Smithson is obvious. Gilbert's remarks produced two other public notices. A week or so later *The Times* observed Smithson's passing and printed the text of his will, and the *Gentleman's Magazine* printed a short obituary. There was no notice from the Royal Institution or in any foreign societies or journals, and once Smithson's estate had been distributed to his heirs, that was the end of it. In 1832, as part of a broader effort to standardize mineral names, the French mineralogist Francois Beudant suggested that zinc carbonate be renamed "smithsonite," but apart from this Smithson's story seemed to be over, and he seemed destined to be forgotten.

Then in 1835 Smithson's nephew, who had inherited his estate, suddenly died. More important, he died without having fathered any children. This activated the final clause in Smithson's will, which was that in the absence of heirs, his estate should go to the United States to found "at Washington, under the name of the Smithsonian Institution, an establishment for the increase and diffusion of knowledge among men."

James Smithson is correctly credited for founding the Smithsonian, but what is rarely mentioned is the remarkably passive way in which he did it. He never wrote another word about his wishes for the new institution, and he does not appear to have discussed it with anyone—not with a lawyer, since he wrote the will himself, and not with anyone in the US government. Indeed, when President Andrew Jackson first learned of the bequest, he thought the United States could not legally accept it.

In 1838, Smithson's estate finally arrived in New York harbor. Most of it came in the form of gold coins, with a total value of over $500,000. This was a huge amount at the time, yet Smithson's only instructions were to make "an establishment for the increase and diffusion of knowledge among men." What exactly did he mean? A congressional committee studied the question but found no clear answer. One of the possibilities was that Smithson had been thinking about worker education, and several of his articles do suggest an interest in this, but it was not realized at the time. Besides, it had been twelve years since Smithson wrote his will, and in that time the situation had changed. The worker-education movement was growing in many American cities, and the needs of workers were beginning to be addressed. Congress did not see this as a priority.

The debate about what to do with Smithson's money dragged on, and many proposals were put forward. A university, a great library, and an astronomical observatory were among them, although none of these had a specific connection to Smithson or his interests. With the lack of clear instruction for its use, his estate was in danger of being used to fund someone's pet project.

The need to parse Smithson's phrase created the first wave of interest in him and his science, and in 1844 American chemist Walter R. Johnson presented an influential paper to the National Institute for the Promotion of Science. "In the many notices of Mr. Smithson's bequest, and plans for establishing an institution on its basis," he observed, "no succinct account has . . . been offered of the scientific pursuits of Mr. Smithson himself,—a very material omission." Searching for clues, Johnson read Smithson's scientific articles, along with some two hundred unpublished articles and hundreds of notes and memoranda that had come with his estate. He characterized these unpublished works as being "of a cyclopedical character. Many of these are connected with general subjects of history—the arts—language—rural pursuits—gardening—the construction of buildings, and kindred topics, such as are likely to occupy the thoughts and to constitute the reading of a gentleman of extensive acquirements and liberal views."

Johnson ended his paper with a question: "What would have been the purposes of an institution founded by Smithson in his lifetime?" And his answer was, "researches to 'increase' positive knowledge, and publication to 'diffuse' and make that knowledge available to mankind." In Johnson's view, this new Smithsonian Institution should reflect the interests of its founder, and since Smithson had possessed the almost unlimited interests of a gentleman of the late Enlightenment, the new institution should have a similar range. In 1846, after eight years of sometimes heated debate, Congress finally agreed and opted for the widest possible interpretation of Smithson's wishes. Without interpreting it any further, the increase and diffusion of knowledge became the new institution's mission—along with the tacit assumption that science would always play an important part.

In the early days of the Smithsonian, Johnson's view of Smithson's science was generally followed. As Johnson put it, "his was not the character of a mere amateur of science. He was an active and industrious laborer in the most interesting and important branch of research—mineral chemistry."

By the 1870s, the success of the Smithsonian had created a desire to know more about its founder. It had been almost fifty years since Smithson's death, and the lack of even a basic biography was becoming an embarrassment. William Rhees, chief clerk of the Smithsonian, was given the assignment of putting one together. Rhees was neither a scientist nor a historian, and he lacked much of the information that had been available to Johnson. Smithson's unpublished articles, along with hundreds of notes and memoranda, his tools, and his mineral collection, had all been destroyed in a fire in 1865. All that was left were his published articles, and these became the starting point for a new assessment of Smithson's science, which was published in 1879 as *The Scientific Writings of James Smithson*. This book, as mentioned in the introduction, concluded that Smithson had been—at best—a second-tier figure of few notable accomplishments. This assessment had the weight of Smithson's own institution behind it, and throughout the twentieth century it was accepted with little question.

But the inadequacy, even wrongness, of this view has become apparent in recent years. The idea that he was an amateur or scientific dabbler does not agree with his having been on the Royal Society's governing council, and it cannot explain the obvious respect he received from the best scientific figures of his time. A new assessment of Smithson is needed, and I hope we have made a start on it in these pages.

James Smithson once wrote about his desire to be remembered. "The best blood of England flows in my veins; on my father's side I am a Northumberland, on my mother's I am related to kings, but this avails me not. My name shall live in the memory of man when the titles of the Northumberlands and the Percys are extinct and forgotten." The context here is his illegitimate birth. Smithson's tone is defiant, because he is determined to be remembered for his own accomplishments, not those of his ancestors.

But if fame was something he desired, he did little to acquire it. Smithson's discoveries and innovations were important—more important than previously thought—and they made him well known and respected in the scientific community, but not outside it. The kind of science Smithson practiced, although of considerable interest, was unlikely to bring him fame. He also had a reserved nature, and apart from his published articles, he did little to call attention to himself. There is no evidence of him ever giving a speech, teaching a class, editing a journal, or writing a book, and on the few occasions that his work was criticized, he never chose to respond.

Smithson's lifestyle may also have worked against him. He was social and had many friends, but he traveled frequently, often for years at a time. He never married, never owned a house, and seldom lived in the same place for more than a few years. Early in his career he was active in the Royal Society, but he broke with it after a dispute (although he officially remained a member). That was around 1817, and his interactions with the Royal Institution seem to have ended around the same time. Smithson spent most of his remaining years in Paris without any active scientific affiliations, which may explain the lack of French obituaries or the customary biographical notes.

Smithson was almost forgotten when news of his gift captured the public imagination and made him famous. The charter that Congress passed in 1846 put him into a new category: institution founder. Disposing of his estate the way he did was a gamble, with little likelihood of success. Yet when Smithson wrote his will he was a gambler, using the excitement of wagering like a tonic to distract him from his pain. Was that what motivated him to put that last clause in his will? Was Smithson placing a bet on the future?

If so, it was a wise one. Today, the Smithsonian Institution consists of nineteen museums, twenty-one libraries, nine research centers, and the National Zoo, together with facilities and affiliations around the globe. It is the largest museum complex in the world and houses an estimated 155 million objects, works of art, and specimens. Its buildings line both sides of the National Mall in Washington, DC, and James Smithson's body now resides in a marble crypt in the Smithsonian Castle, where he is honored—and remembered—for both his scientific achievements and his vision.

———————+———————

NOTES

INTRODUCTION

1. "the National Mall in Washington, DC." Heather Ewing, *The Lost World of James Smithson* (New York: Bloomsbury, 2007), 324.

2. "ravaged the Smithsonian Castle in 1865." *Annual Report of the Board of Regents of the Smithsonian Institution for 1857* (Washington, DC, 1858), 35.

2. "and privately circulated in England." *Nature* 22 (June 3, 1880): 110. The history of early attempts to compile a Smithson biography is summarized in William L. Bird, "A Suggestion Concerning James Smithson's Concept of 'Increase and Diffusion,'" *Technology and Culture* 24 (1983): 246–249.

2. "'he was, nevertheless, no unworthy seeker.'" James Smithson, *The Scientific Writings of James Smithson*, ed. William J. Rhees (Washington, DC, 1879), 143.

3. "'a thorough and an indefatigable one.'" Samuel Pierpont Langley, "James Smithson," *Scientific American Supplement* 22 (February 6, 1904), 23495; "The Centenary of James Smithson," *Science* 70, no. 1801 (July 1929): 8.

3. "'useful to future research chemists.'" Leonard Carmichael and J. C. Long, *James Smithson and the Smithsonian Story* (New York: Putnam, 1965), 115.

1. THE LONG ROAD TO STAFFA

7. "young boy's sense of isolation." This narrative draws from Heather Ewing, *The Lost World of James Smithson* (New York: Bloomsbury, 2007), 19–48.

8. "dedicated . . . to 'Sir Hugh Smithson.'" James Smithson, "On a Saline Substance from Mount Vesuvius," Royal Society Archives, L&P/30/13; William Lewis, *A Course of Practical Chemistry*, 5th ed. (London, 1746). It appears that Lewis and Smithson's father were in a business arrangement.

8. "'other resident members of the University.'" Davies Gilbert, "Address Delivered before the Royal Society," *Abstracts of the Papers Printed in the Philosophical Transactions of the Royal Society of London* 2 (1830–37): 6–12.

8. "one of whom was Smithson." Albert Edward Musson and Eric Robinson, *Science and Technology in the Industrial Revolution* (London: Routledge, 1969), 176. The other three chemists were William Higgins, Davies Giddy and Thomas Beddoes. Peter J. T. Morris, "The Eighteenth Century: Chemistry Allied to Anatomy," *Chemistry at Oxford: A History from 1600 to 2005*, ed. Robert J. P. Williams, Allan Chapman, and John S. Rowlinson (London: RSC Publishing, 2009), 65.

9. "seventy years earlier." "Newtonian" chemistry, introduced at Oxford around 1704, was still being taught at Oxford up to the time that Wall arrived. Williams, Chapman, and Rowlinson, *Chemistry at Oxford*, 71.

9. "up-to-date laboratory." *Correspondence of Joseph Black*, ed. Jean Jones and Robert G. W. Anderson (London: Taylor & Francis, 2012), 1:441, 466; Martin Wall, *Dissertations on Select Subjects in Chemistry and Medicine* (1783), vii; Williams, Chapman, and Rowlinson, *Chemistry at Oxford*, 6–7; A. V. Simcock, *The Ashmolean Museum and Oxford Science, 1683–1983* (Oxford: Oxford University Press, 1984), 8–9.

9. "unexpected appeal to the divinity students." Margaret Evans, ed., *Letters of Richard Radcliffe and John James of Queen's College, Oxford 1755–83* (Oxford: Oxford University Press, 1888), 164–178; Musson and Robinson, *Science and Technology in the Industrial Revolution*, 175.

11. "he probably had Hornsby to thank for this." For example, Smithson's membership proposal for the Royal Society described him as "a Gentleman well versed in various branches of Natural Philosophy & particularly in Chymestry and Mineralogy," EC/1787/03, Royal Society.

11. "all the known gases." The gases studied in Hornsby's course were primarily hydrogen, oxygen, carbon monoxide, and nitrogen.

11. "building a mineral collection." Musson and Robinson, *Science and Technology in the Industrial Revolution*, 176; J. A. Bennett, S. A. Johnston, A. V. Simcock, *Solomon's House in Oxford: New Finds from the First Museum* (Oxford, 2000), 19–20. Wall furnished Thomson with a letter of introduction to Joseph Black: *Correspondence of Joseph Black*, 465–466. Matthew D. Eddy, *The Language of Mineralogy: John Walker, Chemistry and the Edinburgh Medical School, 1750–1800* (New York: Taylor & Francis, 2008), 105, 248.

11. "in the Inner Hebrides called Staffa." This observation came from a 2014 conversation with Tony Simcock, Librarian at Oxford's History of Science Museum, who noted that the wait for an MA to receive an appointment was typically about two years. Barthélemy Faujas de Saint-Fond, *Travels in England, Scotland, and the Hebrides* (London, 1799), 2:61; J. E. Stock, *Memoirs of the Life of Thomas Beddoes, M.D.* (Bristol, 1811), 244; Hugh Torrens, "The Geological Work of Gregory Watt," in *The Origins of Geology in Italy*, ed. Gian Battista Vai and W. Glen, and E. Caldwell (Washington, DC: Geological Society of America, 2006), 183; R. T. Gunther, "Dr. William Thomson, F.R.S., a Forgotten English Mineralogist," *Nature* 143 (1939): 667.

11. "including a visit to Staffa." Joseph Black, *Correspondence*, 697, 727.

12. "travel to distant locations was difficult and expensive." Barthélemy Faujas de Saint-Fond, *Recherches sur les volcans éteints du Vivarais et du Velay* (Paris: Nyon, 1778).

12. "lent it a distinct air of celebrity." This was in March 1784. Ewing, *Lost World*, 71.

12. "'with pleasure, into our party.'" Faujas, *Travels in England, Scotland and the Hebrides*, 1:127.

12. "and one each for Faujas and Smithson." Ibid., 1:281.

13. "nearby island made of stone columns." Joseph Banks, "Account of Staffa," in Thomas Pennant, *A Tour in Scotland and Voyage to the Hebrides, MDCCLXXII* (London: Benj. White, 1790), 299–308.

15. "philosophical debate among gentlemen." Uno von Troil, "Letter XXV, From Professor Bergman to Dr. Troil," *Letters on Iceland* (London: J. Robson, 1780), 366.

15. "Scotland was largely volcanic." Martin Wall, *A Syllabus of a Course of Lectures* (Oxford, 1782), 25; Barthélemy Faujas de Saint-Fond, *Minéralogie des volcans, ou Description de toutes les substances produites our rejetées par les feux souterrains* (Paris, 1784).

16. "the excellent roads between London and Edinburgh." Faujas, *Travels in England, Scotland and the Hebrides*, 1:126; Ewing, *Lost World*, 78.

16. "a longer visit on the way back." Faujas, *Travels in England, Scotland and the Hebrides*, 1:133, 222; Ewing, *Lost World*, 82–85.

16. "founded in the United States is noteworthy." John Anderson, *Institutes of Physics* (Glasgow, 1777). Some of Anderson's teaching instruments survive at the University of Strathclyde. "Anderson, John (1726–1796)," *Dictionary of National Biography* (London, 1885–1900). Smithson would have known about the "Andersonian" from its first natural philosophy professor, Thomas Garnett, who later moved to London and was active in the Royal Institution, of which Smithson was a member. George Birkbeck, who replaced Garnett at the Andersonian, also later moved to London and was active as a chemist and in worker education.

17. "all was not well within the group." Faujas, *Travels in England, Scotland and the Hebrides*, 1:240–258; Ewing, *Lost World*, 70, 87.

19. "but otherwise uninjured." Smithson's journal of the expedition, along with almost all his other papers, was destroyed in the 1865 fire in the Smithsonian Castle, but a few passages from his time on Staffa were saved and copied in Walter R. Johnson, "Memoir on the Scientific Character and Researches of James Smithson," in *The Scientific Writings of James Smithson*, ed. William J. Rhees (Washington, DC, 1879), 140. Faujas, *Travels in England, Scotland and the Hebrides*, 2:22–23.

19. "he would make good use of them in an article." James Smithson, "On the Composition of Zeolite," *Philosophical Transactions* 101 (1811): 171.

19. "sank off the Scottish coast." Faujas, *Travels in England, Scotland and the Hebrides*, 2:52, 59–62, 186.

19. "'most famed after the Collumns.'" Charles Greville to William Hamilton, n.d. [Oct. 1784], in *The Hamilton and Nelson Papers* (London, 1893–94), 91–92.

20. "no evidence in print that they had ever existed." J. S. Herman to J. Black, quoted in Alec Livingstone, *Minerals of Scotland: Past and Present* (Edinburgh: National Museums of Scotland, 2002).

20. "elicited little scientific interest." Faujas, *Travels in England, Scotland and the Hebrides*, 2:58, For a contemporary assessment of Staffa's importance, see John Macculloch, *A Description of the Western Islands of Scotland* (London, 1819), 2:18–22.

21. "dramatic opening of Fingal's Cave." Adrienne L. Kaeppler, *Holophusicon: The Leverian Museum* (Honolulu: Bishop Museum Press, 2011), 50–52; Richard D. Altick, *The Shows of London* (Cambridge, MA: Harvard University Press, 1978), 235–246; Ralph O'Connor, *The Earth on Show* (Chicago: University of Chicago Press, 2007), 37; National History Museum, http://www.nhm.ac.uk/visit.html (last accessed September 16, 2019).

2. EDINBURGH, LONDON, AND PARIS
————

24. "how influential he was." V. A. Eyles, "The Evolution of a Chemist," *Annals of Science* 19 (1963): 157–158; W. P. Doyle, "Joseph Black (1728–1799)," http://www.chem.ed.ac.uk/about-us/history-school/professors/joseph-black (last accessed September 15, 2019).

24. "a correspondence with Black that would continue for many years." William Thomson to Joseph Black, 28 August 1784, and Lewis Macie to Black, 27 February 1785, *Correspondence of Joseph Black*, ed. Robert G. W. Anderson and Jean Jones (Farnham: Ashgate Publishing, 2012), 1:727 and 795–796.

24. "a project he had been working on for more than twenty years." Smithson was in Edinburgh in October 1784; see Barthélemy Faujas de Saint-Fond, *Travels in England, Scotland, and the Hebrides* (London, 1799), 2:230. A full account of Hutton's theory was read at the March 7, 1785, and April 4, 1785, meetings of the Royal Society of Edinburgh. It was published in 1788: James Hutton, "Theory of the Earth; or an Investigation of the Laws observable in the Composition, Dissolution, and Restoration of Land upon the Globe," *Transactions of the Royal Society of Edinburgh* (1788), 1:209–304.

24. "sites hundreds of miles away." Faujas, *Travels in England, Scotland, and the Hebrides*, 2:230; Jean Jones, "The Geological Collection of James Hutton," *Annals of Science* 41 (1984): 235.

25. "'may venture to follow.'" John Playfair, "Account of the Late Dr. James Hutton," *Transactions of the Royal Society of Edinburgh* 5 (1805): 39–99.

25. "mutual affection and respect." For example: "Joseph Black to James Macie [Smithson], Sept. 18, 1790," *Correspondence*, 2:1098, and "Joseph Black to Charles Greville, Oct. 12, 1790," ibid., 2:1100.

26. "nowhere else in the world." Michael P. Cooper, *Robbing the Sparry Garniture* (Tucson, AZ: The Mineralogical Record, 2006), 59.

26. "Black did not have enough to share." *Correspondence*, 1:796. Smithson was also searching for a greenish crystal that he called "phosphorated lead." Black told Smithson to contact a Mr. Stirling, manager of the Scots Mining Company, but Smithson reported that he had not been there. "Lewis Macie to Black; London, 27 February 1785," *Correspondence*, 1:795.

27. "power was disconnected." William Withering, "Experiments and Observations on the Terra Ponderosa," *Philosophical Transactions* 74 (1784): 293–301. Black had also performed an extensive series of experiments on all these materials five years earlier and seems to have already discovered everything that Withering found—though, typically, without publishing a word of it. *Correspondence*, 1:382.

27. "to visit a salt mine." Black, *Correspondence*, 1:697, 733, 763, 795; Walter R. Johnson, "Memoir on the Scientific Character and Researches of James Smithson," in *The Scientific Writings of James Smithson*, ed. William J. Rhees (Washington, DC, 1879), 140; Heather Ewing, *The Lost World of James Smithson* (New York: Bloomsbury, 2007), 94–95.

27. "accompany Smithson to the meeting." Ewing, *Lost World*, 77, 109.

27. "an unqualified success." There is an intriguing suggestion that he may have been able to simply buy his sample from one of the city's mineral dealers, who had misidentified it. Joseph Black once wrote that "many fossilists [mineral dealers] may have it [Terra Ponderosa Aërata] in England under the name of Scotch Zeolite." *Correspondence*, 1:437.

28. "Smithson graduated from Oxford." Christa Jungnickel and Russell McCormmach, *Cavendish: The Experimental Life* (Malden, MA: Blackwell, 2001), 427; Trevor Levere and Gerard L'E Turner, *Discussing Chemistry and Steam: The Minutes of a Coffee House Philosophical Society 1780-1787* (Oxford: Oxford University Press, 2002), 162.

28. "could make social interaction with him difficult." J. Z. Fullmer, "Davy's Sketches of His Contemporaries," *Chymia* 12 (1966): 133.

29. "avoided contact with them, and never married." Henry Lord Brougham, "Cavendish," in his *Lives of Men of Letters and Science Who Flourished in the Time of George III* (London: Charles Knight, 1845), 1:444-445.

29. "'but you may set him going.'" Quotes about Cavendish come from George Wilson, *The Life of the Hon. Henry Cavendish* (London, 1851), 166-170; Jungnickel and McCormmach, *Cavendish*, 303-309.

29. "'seemed to take great pleasure in his company.'" *Minute Book, Royal Society Club*, RS, 8; Jungnickel and McCormmach, *Cavendish*, 304; "4 January 1794, from Charles Blagden to Georgiana, Duchess of Devonshire," Chatsworth MS CS5/1202. Quoted in Ewing, *Lost World*, 172.

29. "monitored by a select group of witnesses." Henry Cavendish, "Experiments on Air," *Philosophical Transactions* 75 (1785): 372; "On the Conversion of a Mixture of Dephlogisticated and Phlogisticated Air into Nitrous Acid by the Electric Spark," *Philosophical Transactions* 78 (1788): 261.

30. "he was just twenty-two years old." Cavendish, along with Richard Kirwan and Charles Blagden, Joseph Banks' assistant, all signed the certificate proposing Smithson for membership. For a more detailed discussion of that election, see Ewing, *Lost World*, 118-120.

30. "could have legitimately been elected a Fellow." "Elections," Royal Society web site, https://royalsociety.org/about-us/fellowship/election (last accessed September 16, 2019).

30. "'a very useful & valuable member.'" Royal Society, *Certificates*: "James Lewis Macie," 5 (19 April 1787), EC/1787/03.

30. "'without acquiring some new ideas.'" "4 January 1794, from Charles," Chatsworth MS CS5/1202. Quoted in Ewing, *Lost World*, 172.

30. "shortly thereafter Smithson left for France." From Banks's letter of introduction to Lavoisier (1788), quoted in Ewing, *Lost World*, 125; Cavendish, "On the Conversion of a Mixture of Dephlogisticated Air," 261-276.

31. "the mineralogist René Just Haüy." "Joseph Banks to Antoine Lavoisier, April 8, 1788," Papiers Lavoisier, Archives du Comte de Chabrol, Archives de l'Académie des Sciences, Paris.

31. "would continue for more than three decades." René Just Haüy, *Essai d'une théorie sur la structure des crystaux: Appliquée a plusieurs genres de substances crystallisées* (Paris, 1784).

32. "without ever having used them." "Letter from William Thomson, Oxford, dated 16 Dec. 1789," Cornwall Record Office: CRO J3/2/33. Quoted in Ewing, *Lost World*, 172.

32. "into broader scientific notice." See, for example, Jacques Louis Bournon, *Traité de minéralogie*, vol. 2 (London: William Phillips, 1808); "Macie letter to Greville, sent Jan. 1, 1792," British Library, MS41199 f.82; "Macie letter to Charles Greville" n.d., British Library, Hamilton and Greville Papers, vol. IV, f.164–65, written after Oct. 6, 1790; "Smithson letter to Charles Greville," Hamilton and Greville Papers, vol. IV, f.166, written after July 1789. Macie is mentioned in a review of Haüy's *De la structure, considerée comme caractere distinctif des minéraux* in *The British Critic* 5 (1795): 310; Haüy, *Traité de minéralogie* (Paris, 1801), 2:262; Haüy, *Comparative Table of Results of Crystallography and Chemical Analyses Relative to the Classification of Minerals* (Paris, 1809), n. 154; Haüy, *Traité de minéralogie* (Paris, 1822–23), 312.

32. "engaged him completely." René Just Haüy, "Des observations sur la vertu électrique que plusieurs minéraux acquièrent à l'aide de la chaleur," *Choix de mémoires sur divers objets d'histoire naturelle* (Paris, 1792).

3. TABASHEER

33. "he suspected of containing more." Patrick Russell, "An Account of the Tabasheer," *Philosophical Transactions* 80 (1790): 273–283.

33. "samples in the bags." James Macie, "An Account of Some Chemical Experiments on Tabasheer," *Philosophical Transactions* 81 (1791): 368–388.

34. "'the bodies of vegetables themselves.'" The introduction of the three "kingdoms" is generally credited to the French chemist Nicolas Lemery in his textbook *Cours de chymie* (Paris, 1675). See William Lewis, *The Chemical Works of Caspar Neumann*, 2nd ed. (London, 1773), 408; see also Ursula Klein and Wolfgang Lefèvre, *Materials in Eighteenth-Century Science: A Historical Ontology* (Cambridge, MA: MIT Press, 2007), 241–242.

34. "organic/inorganic distinction." The philosophical distinction between organic and inorganic was finally put to rest in 1828 when the German chemist Friedrich Wohler synthesized the organic compound urea from the purely inorganic ammonium cyanate.

34. "'are now under chemical trial.'" Charles Greville to Joseph Banks, n.d., BL Add MS 33982.F.238. Blagden Papers 7.322, RS, quoted in Heather Ewing, *The Lost World of James Smithson* (New York: Bloomsbury, 2007), 142. The announcement was contained in a postscript, dated July 16, 1790, to Russell, "An Account of the Tabasheer," 283. Although Smithson's analysis did not address the medical uses of tabasheer, this was also a topic of interest. See *Medical Commentaries for the Year MDCCXCI* (Edinburgh, 1792), 6:103–113.

35. "ensured a common methodology." See, for example, Torbern Bergman, *Outlines of Mineralogy* (Birmingham, 1783), and Axel Frederic Cronstedt, *An Essay towards a System of Mineralogy* (London, 1788). Smithson appears to have consulted both these works.

35. "to reach a conclusion." Macie, "An Account of Some Chemical Experiments on Tabasheer," 368–388; unless otherwise noted, quotations in the following discussion are drawn from this article.

36. "mostly of 'siliceous earth.'" Use of the term "earth" was common in Smithson's time and referred to the four classes of minerals, which were metals, combustible materials (like coal), "salts" (which included most of the commercially valuable minerals), and "earths" (which was the least understood class and included, among many others, the oxides of calcium, aluminum, and silica). The terms "siliceous," "siliceous earth," and "silex" were used interchangeably by

Smithson and correspond well with the modern "silica." They are all used interchangeably in the current work.

36. "often included in mineralogical tests." Physical descriptions were also used to classify minerals. See, for example, Abraham Werner, *A Treatise on the External Characters of Fossils* (Dublin, 1805).

36. "not present in the form of flint." J. H. Pott, *Lithogéognosie ou examen chymique des pierres et des terres en général* (Paris, 1746); William Nicholson, "Analysis," *A Dictionary of Practical and Theoretical Chemistry* (London, 1809). This effect was well known from the use of flintlock firearms.

37. "thus did not contain fluorite." Bergman, *Outlines of Mineralogy*, 46, item #96.

37. "the purest form of siliceous earth." Richard Kirwan, *Elements of Mineralogy*, 2nd ed. (London, 1794), 1:2, 10.

38. "tabasheer's identifying characteristics." For more on this see David Brewster, "On the Optical and Physical Properties of Tabasheer," *Philosophical Transactions* 1 (1819): 283–299.

38. "particularly siliceous earth." Russell, "An Account of the Tabasheer," 274.

38. "when it evaporated." See for example, Kirwan, *Elements of Mineralogy*, 10; Torbern Bergman, "On the Effects of Fire, Both at the Volcanos and the Hot Springs; and also of the Basalts," in *Letters on Iceland* (London, 1780), 345–347.

38. "finding any white film." Smithson's "soft" water for this test was likely produced by treating his distilled water with "limewater" (calcium hydroxide).

38. "a form of siliceous earth." For more on storing these natural materials, see James Watt, "On a New Method of Preparing a Test Liquor to Shew the Presence of Acids and Alkalies in Chemical Mixtures," *Philosophical Transactions* 74 (1784): 419–422.

40. "completely resistant to combustion." See, for example, experiment 4 in Nicholson, "Analysis," quoted in J. R. Partington, *A Short History of Chemistry*, 3rd ed. (London: Macmillan, 1957), 74.

41. "'more cracked by the fire.'" Cronstedt, *An Essay towards a System of Mineralogy*, 1:120–122, test 4. Smithson's copy of this book is in the Smithsonian Libraries, QE362.C7613 1788.

41. "it did not melt." Nicolas-Théodore de Saussure, "Lettre de M. de Saussure de Genève, à M. l'Abbé Mongez le jeune, sur l'usage du chalumeau," *Observations sur la physique* 26 (1785): 409. Smithson's citation of this article used *Journal de Physique*.

41. "listed the characteristics of 'terra silicea' (silica)." Bergman, *Outlines*, 60.

41. "by boiling it in acids." See, for example, Cronstedt, *An Essay towards a System of Mineralogy*, 1:120–122; Bergman, *Outlines*, 59–60.

41. "reduced in weight." Cronstedt, *An Essay towards a System of Mineralogy*, 1:120–122.

42. "The blowpipe flame." George J. Brush, *Manual of Determinative Mineralogy* (New York: John Wiley & Son, 1875), 6.

42. "an impressive feat." This also demonstrated the absence of "lime" (calcium oxide). Pierre-Joseph Macquer, "Sugar," *Dictionary of Chemistry* (London, 1777); Macquer notes that there is a "powerful affinity of the acid of sugar with calcareous earth."

43. "dangerous a material to work with." Bergman, *Outlines*, 17, 60; Cronstedt, *An Essay towards a System of Mineralogy*, 1:121.

43. "test for its presence." The mild form of each alkali (carbonate) was known to be converted into the caustic form (hydroxide) by treatment with slaked lime. "It [silex] may be dissolved by the fixed alkali [either potash or soda], both in the dry and wet way," Cronstedt, *An Essay towards a System of Mineralogy*, 1:120–122.

45. "'a pellucid [transparent] glass.'" Torbern Bergman, *Physical and Chemical Essays* (London, 1784), 2:484–485.

46. "'the humidity of the atmosphere.'" Cronstedt, *An Essay towards a System of Mineralogy*, 1:120–122.

47. "the presence of silica." Ibid.; Kirwan, *Elements of Mineralogy*, 2.

47. "his overall conclusion." Frederick Hutton Getman, *The Elements of Blowpipe Analysis* (New York: Macmillan, 1899), 18. M. I. A. Chaptal, *Elements of Chemistry* (London, 1791), 3:341–342.

47. "'sensible diminution of size.'" Sir William Crookes, *A Manual of Practical Assaying* (New York: John Wiley and Sons, 1881), 221.

48. "having the right proportions." "With oxides, more especially those of lead, it combines by fusion, and forms glass of a dense texture and strong refractive power." Nicholson; "Earth: Silex," *Dictionary*.

48. "no more tabasheer left to use." Smithson appears to have used up all but one of the pieces of Hyderabad tabasheer, and that one he returned to the Royal Society. In 1819, David Brewster made an extensive study the optical properties of tabasheer and reported receiving that piece from the society in *The Edinburgh Philosophical Journal* 1 (1819): 149.

48. "'the product of a vegetable.'" See, for example, Pierre Macquer, *Dictionnaire de chimie* (Paris, 1766), as discussed in Klein and Lefevre, *Materials in Eighteenth-Century Science*, 247.

49. "'the size of half a pea.'" Macie, "Some Chemical Experiments on Tabasheer," 387. Curiously, the pebble Banks found was much darker and harder than ordinary tabasheer and was found, on analysis, to contain iron—an anomalous finding not encountered by any subsequent investigators.

49. "'which M. Macé has analysed.'" James Edward Smith, *An Introduction to Physiological and Systematical Botany*, 2nd ed. (London, 1809), 75. Macie is cited on page 76; see also citations of Macie in Richard Duppa, *The Classes and Orders of the Linnaean System of Botany* (London, 1816), 3:534, and in *Annals of Agriculture and Other Useful Arts* 23 (1795): 110. "Letter from M. Humboldt to C. Delambres, one of the perpetual Secretaries of the National Institute," *Philosophical Magazine* 16 (1803): 172; this letter was dated "Lima, Nov. 25, 1802" and was originally published in *Annales de Museum d'Histoire Naturelle* 8.

50. "entangled with national pride." Vauquelin, *Memoires de l'Institut* 6 (1806): 382; Alexander von Humboldt, "Remarks on the Natural Family of the Grasses," *Journal of Science and the Arts* 5 (1818): 50.

50. "the accuracy of Smithson's analysis." Edward Turner, "Chemical Examination of Tabasheer," *The Edinburgh Journal of Science* 8 (1828): 335–336; Thomas Thomson, "Chemical Analysis of Tabasheer," *Arcana of Science and Art: or an Annual Register of Useful Inventions and Improvements, Discoveries and New Facts* (London, 1837), 152.

51. "'refer the analysis of the tabasheer.'" Richard Kirwan, *Geological Essays* (London, 1799), 116–117, 467; Jacques François Demachy, *Instituts de chymie, ou, Principes élémentaires de cette science, présentés sous un nouveau jour* (Paris, 1766). In Smithson's time this idea was being developed by the French naturalist Jean Baptiste Lamarck, who published a more refined

version, *Hydrogéologie, ou, Recherches sur l'influence qu'ont les eaux sur la surface du globe terrestre* (Paris, 1802). See also Albert V. Carozzi, "Lamarck's Theory of the Earth: Hydrogeologie," *Isis* 55 (1964): 293–307; "French National Institute," *Philosophical Magazine* 22 (1805): 176–177.

4. CALAMINE

53. "Thomson was in serious trouble." William Thomson to George Paton, 25 Sept 1790, Paton MSS, quoted in: Hugh S. Torrens, "Thomson, William," *Oxford Dictionary of National Biography*, https://www.oxforddnb.com (last accessed March 17, 2015); EUL Gen 873/II/158–159; "Minutes and Register of Convocation," Oxford University archives, quoted in Hugh S. Torrens, "The Geological Work of Gregory Watt," in *The Origins of Geology in Italy*, ed. Gian Battista Vai, W. Glen, and E. Caldwell (Washington, DC: Geological Society of America, 2006), 183–184.

53. "Thomson's rise in English science had been as impressive as Smithson's." Thomson was elected Lee's Reader in the Anatomy School, April 1785; named physician at the Radcliffe Infirmary, June 12, 1786; and elected to the Royal Society, March 16, 1786. Heather Ewing, *The Lost World of James Smithson* (New York: Bloomsbury, 2007), 185–186.

54. "banished from the university." EUL Gen 873/II/158–189.

54. "he felt obliged to attend." H. R. V. Fox, *Further Memoirs of the Whig Party, 1807–1821* (London: Murray, 1905), 340; quoted in Torrens, "The Geological Work of Gregory Watt," 183–184.

54. "was effectively over." In an unpublished timeline, Heather Ewing quotes entries in records of the Royal Society for November 11, 1790 and the Society for Promoting Natural History for December 20, 1790, in which Thomson's resignation letters are read to the members. My thanks to her for access to this document.

54. "to have used his services." Dr. R. T. Gunther, "Dr. William Thomson, F.R.S., a Forgotten English Mineralogist, 1761–c. 1806," *Nature* 143 (1939): 667–669.

54. "'possibly at Naples.'" Ewing, *Lost World*, 162, 194; Smithson to Greville, January 1, 1792, British Library, MS 41199, f.82, my emphasis.

55. "'but monks and convents.'" Ibid.

55. "'[with] his teeth the lion eats him'" and "'the final death stroke.'" Ibid.; "9 May, 1792, James Macie to Davies Giddy," Cornwall PRO; this letter is copied in the *Smithsonian Institution Annual Report, 1884*, 5–6.

56. "'men in a collected state.'" British Library, Add MS 51822 ff. 54–5—Smithson to Holland, December 3, 1801 (written from Dover); Ewing, *Lost World*, 385–386n85, 391n51.

56. "the size of a small suitcase." W. A. Smeaton, "The Portable Chemical Laboratories of Guyton de Morveau, Cronstedt and Gottling," *Ambix* 13 (1966): 84–91.

57. "samples from Mount Vesuvius." Torrens, "The Geological Work of Gregory Watt," 184.

57. "more complete analysis of it." Guglielmo [William] Thomson, *Breve notizia di un viaggiatore sulle incrostazioni silicee termali d'Italia* (Rome, 1795); it was also reprinted in French, in the first issue of the journal *Bibliothèque Britannique* (1796). Smithson, "On a Saline Substance from Mount Vesuvius," *Philosophical Transactions* 103 (1813): 257.

57. "perhaps the foremost German chemist of his time." Ewing, *Lost World*, 186–191.

58. "an insider in the English scientific community." Minutes of the RI, vol. 1, 123. Charles Greville, William Hamilton, and John Hawkins were all members of the Royal Society, JBC 37. Smithson was sworn into the Royal Society Council on January 15, 1801, R. S. Council Minutes 1782–1810, vol. 7.

58. "officially granted in February 1801." Ewing, Lost World, 204–208; PRO, H. O. Warrant Book, 1801–1802, vol. 48; official notice of Smithson's name change appeared in the London Gazette; Ewing, Lost World, 202.

58. "than either metal by itself." James Smithson, "A Chemical Analysis of Some Calamines," Philosophical Transactions 93 (1803): 12–28.

60. "to establish an English brass industry." Hamilton, The English Brass and Copper Industries to 1800 (London: Cass, 1967), 9–20.

60. "topic of special interest in England." "Jamestown Special Issue," Rittenhouse 21, no. 66 (2007): 65–125.

61. "'by Fumigation with Calamine.'" J. A. Bennett, S. A. Johnston, A. V. Simcock, Solomon's House in Oxford: New Finds from the First Museum (Oxford: Museum of the History of Science, 2000), 32–33; Martin Wall, A Syllabus of a Course of Lectures in Chemistry (Oxford, 1782), 37; Trevor Levere and G. L'E. Turner, Discussing Chemistry and Steam: The Minutes of a Coffee House Philosophical Society 1780-1787 (Oxford: Oxford University Press, 2002), 83.

61. "waiting to be found." Bertrand Pelletier, "D'une substance pierreuse, venant des mines de Fribourg en Brisgaw, désignée par les naturalistes sous le nom de zéolite," Journal de Physique 20, Part II (1782): 420–429.

62. "Smithson set out to explore." Smithson, "Chemical Analysis," 12.

63. "exposed to zinc fumes." Smithson, "Calamines," in The Scientific Writings of James Smithson, ed. William J. Rhees (Washington, DC, 1879), 19; Anonymous, A System of Instruction in the Practical Use of the Blowpipe (New York, 1858), 14–21.

63. "his 'flowers' did exactly that." Edward Salisbury Dana, A Textbook of Mineralogy, rev. ed. (New York: John Wiley, 1932), 373. Note that since the sample was zinc carbonate, the heat would have completely vaporized it.

63. "when tested for it." Smithson, "Calamines," 18. A video of the blowpipe being used to test zinc ores can be seen at https://www.youtube.com/watch?v=UKLrGfiCFt0.

63. "not an assumption that Smithson made." "Calamine [Review]," The Annual Review and History of Literature for 1803 (London, 1804), 2:199; James Nicol, Manual of Mineralogy (Edinburgh: Adam and Charles Black, 1899), 145.

63. "crystals were small and few in number." Smithson, "Calamines," 21.

65. "materials in a given substance." Joseph Proust, "Researches on Copper: on Carbonate of Copper," Annales de Chimie 32 (1799): 26–54.

65. "well into the nineteenth century." Rene Just Haüy, Traité de minéralogie (Paris, 1801). For a detailed discussion of crystallography as the context of Smithson's work, see Stephen T. Irish, "James Smithson on the Calamines: Chemical Combination in Crystals," Ambix 65 (2018): 373; Smithson, "Calamines," 26–27; Seymour H. Mauskopf, "Hauy's Model of Chemical Equivalence: Daltonian Doubts Exhumed," Ambix 17 (1970): 182–191. For more on mathematical simplicity, see John G. Burke, Origins of the Science of Crystals (Berkeley: University of California Press, 1966), 69; for examples of nineteenth-century mathematical simplicity, see Steven Turner,

"Demonstrating Harmony: Some of the Many Devices Used to Produce Lissajous Curves before the Oscilloscope," *Rittenhouse* 11 (1997): 33.

66. "he seemed to be correct." We now know that what Smithson called "carbonic acid" is actually a combination of one atom of carbon with three atoms of oxygen, but in 1803 when Smithson wrote this, chemists had no way to count atoms. The best they could do was to identify the elements in a substance and weigh them.

67. "a compound mineral might be composed." For a summary of the steps of Smithson's analysis, see "Calamine [Review]," *The Monthly Review; or Literary Journal* 42 (1803): 397–398.

67. "analysis of the Bleyberg sample is striking." Smithson, "A Chemical Analysis of Some Calamines," *Philosophical Transactions* 93 (1803): 24.

68. "that was just beginning in England." Smithson, "Calamine," 25–26; Robert Siegfried, "Further Daltonian Doubts," *Isis* 54 (1963): 481.

68. "basis of having heard it read." Humphry Davy, *Journal of the Royal Institution of Great Britain* 1 (1802): 299.

68. "related to the production of brass." *Journal of Natural Philosophy, Chemistry, and the Arts* (1803), 6:74–85; James Smithson, "A Chemical Analysis of Some Calamines," *The Repertory of Arts, Manufactures, and Agriculture* 4 (1804): 254–270.

69. "address in a later publication." "Calamine [Review]," 398. The quotation is taken from a review of Smithson's "Calamine" article in *The Critical Review or Annals of Literature* 1 (1804): 30. Smithson addressed this criticism in "On the Composition of the Compound Sulphuret," *Philosophical Transactions* 98 (1808): 55.

69. "complete article was ever published." Delamétherie, "Extrait d'une letter de M. James Smithson à J. C. Delamétherie," *Journal de Physique* 60 (1805): 179.

69. "the idea of a zinc hydrate." Alexandre Brongniart, *Traité élémentaire de minéralogie* (Paris, 1807), 2:136–140; René-Just Haüy, *Tableau comparative des résultats de la cristallographie et de l'analyse chimique, relativement à la classification des minéraux* (Paris, 1809).

70. "'scrupulously kept the sense of the text.'" M. Smithson, "Mémoire sur les calamines," *Journal des Mine* 28 (1810): 341–362. Extract from *Transactions Philosophiques* (1803), with notes by M. P. Berthier.

70. "'analyses made with great care.'" Smithson (and Berthier), "Mémoire sur les calamines," *Journal des Mines* 167 (November 1810), 358n1. My thanks to Jeff Gorman for this translation.

70. "French studies of the zinc ores." See for example, J. A. H. Lucas, *Tableau méthodique des espèces minérales* (Paris, 1813), 423–424; and Gay-Lussac et Arago, *Annales de Chimie et de Physique* 14 (1820): 394–395.

71. "his particular interest in hydrates." M. H. Klaproth, "Zinc," *Dictionnaire de chimie* (Paris, 1811), 4:536–547; Thomas Thomson, "On the Composition of Blende," *Annals of Philosophy* IV (1814): 89. Thomson credited Smithson with analyzing three of the four known ores of zinc: "Blende," "Hydrous carbonate of zinc," "Anhydrous carbonate," and "Silicated zinc."

71. "'which have since been discovered.'" "Smithson to Berzelius," 14 November 1818, Royal Swedish Academy of Sciences Archives; Jacob Berzelius, "Examination of Some Compounds Which Depend upon Very Weak Affinities," *Edinburgh Philosophical Journal* I (1819): 252. Note that here Berzelius is referring to Smithson's complete article.

5. MINIUM

73. "'became necessary to leave it.'" Smithson to an unidentified Italian, September 4, 1804; letter quoted in Heather Ewing, *The Lost World of James Smithson* (New York: Bloomsbury, 2007), 220.

74. "opposite journey from London to Cassell." Ewing, *Lost World*, 249, reports that at least some of Smithson's correspondence came via his London bankers, who continued to do business on the continent throughout the war. James Smithson, "On the Composition of the Compound Sulphuret from Huel Boys, and an Account of Its Crystals," *Philosophical Transactions* 98 (1808): 55.

74. "'a Discovery of Native Minium (1806).'" British Library, Add MS 51823, ff. 258–259, "Smithson to Lord Holland, Nov. 22, 1805"; James Smithson, "Account of a Discovery of Native Minium," *Philosophical Transactions of the Royal Society* 96 (1806): 267–268. Note that Smithson's letter used the title "Discovery of Native Minium," which was changed by the editor of *Transactions*. Smithson's letter to Banks is dated March 2, 1806; it was read to the Royal Society on April 24, 1806, and published in Part I of the 1806 *Philosophical Transactions*.

75. "the material as being that of minium." Smithson later learned that his minium samples had come from "the lead mines of Breylau in Westphalia," *Philosophical Magazine* 38 (1811): 84.

75. "'factitious [manufactured] minium.'" Today natural minium is regarded as uncommon, but not particularly rare. It forms in conditions of extreme oxidizing, and the best natural specimens now come from Broken Hill in New South Wales, Australia, where they formed as the result of a mine fire.

75. "the sample was a metallic oxide." Smithson wrote: "This quality, of temporarily changing their colour by heat, is common to most, if not all, metallic oxides; the white growing yellow, the yellow red, the red black." James Smithson, "A Chemical Analysis of Some Calamines," *Philosophical Transactions* 93 (1803): 12–28, at 13.

76. "as long as the nitric acid was still potent." See for example, Torbern Bergman, *Outlines of Mineralogy* (Birmingham, 1783), 86.

77. "agents of mineral formation: solvents and heat." This had to do with the ongoing neptunist/vulcanist debate; see, for example, the discussion of minium in James Sowerby, *British Mineralogy* (London, 1809), 3:157.

77. "'cannot subscribe to such an opinion.'" *Journal of Natural Philosophy, Chemistry and the Arts* 16 (1807): 127–128; *Journal de Physique, de Chimie, d'Histoire Naturelle et des Arts* 65 (1807): 365–366; *The Philosophical Magazine* 26 (1806–7): 114–115; "XI.—Account of a Discovery of Native Minium," *Critical Review or Annals of Literature* 10 (1807): 353.

78. "James Sowerby's authoritative *British Mineralogy*." Sowerby, "Plumbum Oxygenizatum," *British Mineralogy*, 155–158.

78. "his discovery of minium in Germany." British Library, Western Manuscripts collection, Hamilton and Greville Papers, vol. IV (ff.365), Add MS 42071, f.164–166, "Smithson to Greville, after 1790." This letter was written between fall 1790 and fall 1792.

78. "did in Germany fifteen years later." Ibid.

78. "identified the letter's author as Smithson Tennant." *The Philosophical Magazine* 24 (1806): 274. Note that this volume is for February–May 1806.

79. "but identified Smithson Tennant as its author." Ibid., 361; *The Scots Magazine and Edinburgh Literary Miscellany* 68 (1806): 328, 895; Sowerby, *British Mineralogy*, 158.

79. "and with no prospect of release." James Smithson, in Germany, to Lord Holland, November 22, 1805; BL Add MS 51823, ff. 258–259. The following narrative draws heavily on Ewing, *Lost World*, 238–258.

79. "'vibrating between existence & the tomb.'" Smithson to Banks, September 18, 1808, Banks Collection, Sutro Library, California; quoted in Ewing, *Lost World*, 245, 250.

80. "Napoleon had ordered his immediate release." The *Institut de France* consisted of five academies, and Delambre was attached to the Académie des Sciences. His formal title was Secrétaire Perpétuel pour les Sciences Mathématiques, or permanent secretary of mathematical sciences. The letter that Delambre wrote to the Minister of War is now in the library of Columbia University. It was translated and discussed in a short monograph: David Eugene Smith, *Delambre & Smithson* (New York: Columbia University, 1934).

6. THE SULPHURET FROM HUEL BOYS

81. "any meetings until the following June." Heather Ewing, *The Lost World of James Smithson* (New York: Bloomsbury, 2007), 255–256.

81. "early years of the Royal Institution." James Smithson, "On the Composition of the Compound Sulphuret from Huel Boys, and an Account of Its Crystals," *Philosophical Transactions* 98 (1808): 55.

82. "some of the advanced courses." Mark I. Grossman, "John Dalton and the London Atomists: William and Bryan Higgins, William Austin, and New Daltonian Doubts about the Origin of the Atomic Theory," *Notes and Records* 68 (2014): 339. Prior to joining the Royal Institution, Young had assisted with lectures at Bryan Higgins's school. See Nicholas A. Hans, *New Trends in Education in the Eighteenth Century* (Malden, MA: Routledge, 1998), 150; see also Joseph Priestley, *Philosophical Empiricism* (London, 1775), 24–26.

82. "a home in the newly founded Royal Institution." Trevor Levere and Gerard L'E Turner, *Discussing Chemistry and Steam: the Minutes of a Coffee House Philosophical Society 1780-1787* (Oxford: Oxford University Press, 2002), 222–223; Gwen Averley, "The 'Social Chemists': English Chemical Societies in the Eighteenth and Early Nineteenth Century," *Ambix* 33 (1986): 105; Bryan Higgins, ed., *Minutes of the Society for Philosophical Experiments and Conversations* (London, 1795), 10. Grossman, "John Dalton and the London Atomists," 347.

82. "after Young joined the Royal Institution." Both William Higgins and William Austin were at Oxford during Smithson's final year. Crista Jungnickel and Russell McCormmach, *Cavendish* (Philadelphia: American Philosophical Society, 1996), 267; G. N. Cantor, "Thomas Young's Lectures at the Royal Institution," *Notes and Records of the Royal Society of London* 25 (1970): 87–112.

83. "'the metals mineralized by Sulphur.'" Joseph Proust, "Sur les mines de cobalt, nickel et autres," *Journal de Physique* 63 (1806): 564–577; Thomas Young, "Review of Proust, 'On Ores, and particularly on those of Cobalt and Nickel,' *Journal de Physique*, November 1806," *Retrospect of Philosophical, Mechanical, Chemical, and Agricultural Discoveries* 3 (1808): 421–425.

83. "binary combination did not apply." Young, "Review of Proust," 425.

83. "'decisive observations on the subject.'" William Wollaston, "On Super-Acid and Sub-Acid Salts," *Philosophical Transactions* 98 (1808): 96, 98.

84. "took them seriously is obvious." Young, "Review of Proust," 425.

84. "developments back in England." Grossman, "John Dalton and the London Atomists," 344; Melvyn C. Usselman, "Multiple Combining Proportions: The Experimental Evidence," in *Instruments and Experimentation in the History of Chemistry*, ed. Frederic Holmes and Trevor Levere (Cambridge, Mass.: MIT Press, 2002), 250; John Dalton, *A New System of Chemical Philosophy*, Part 1 (Manchester, 1808).

84. "French mineralogist Comte de Bournon." Comte de Bournon, "Description of a Triple Sulphuret, of Lead, Antimony, and Copper, from Cornwall," *Philosophical Transactions* 94 (1804): 30–62.

84. "the same issue of the *Transactions*." Charles Hatchett, "Analysis of a Triple Sulphuret, of Lead, Antimony, and Copper, from Cornwall," ibid., 63–69.

85. "as a challenge to his own theory." Frank Greenaway, *John Dalton and the Atom* (London: Heinemann, 1966), 150–152.

86. "system could not be taken seriously." "[Review] On the Compound Sulphuret from Huel Boys. By Mr. James Smithson," *Retrospect* 4 (1809): 168–169; René Just Haüy, *Traité de minéralogie* (Paris, 1801).

86. "'Its nature surprised me.'" Comte de Bournon, "Memoir on the Triple Sulphuret of Lead, Copper, and Antimony, or Endellion," *Journal of Natural Philosophy* 24 (1809): 226.

87. "'simple' proportions are listed on the right." Smithson, "On the Composition of the Compound Sulphuret from Huel Boys, and an account of its Crystals," *Philosophical Transactions* 98 (1808): 57.

87. "he now saw Smithson as his enemy." Smithson, "On the Composition of the Compound Sulphuret from Huel Boys," 55.

87. "'with more respect.'" "[Review] On the Composition of the Compound Sulphuret from Huel Boys, and an Account of Its Crystals, by James Smithson," *The Eclectic Review* 5 (1809): 437; "On the Compound Sulphuret," *Retrospect*, 168–169.

88. "as 'ingenious and probable.'" Bournon, "Memoir," 326.

88. "Smithson's article was evaluated." Thomas Thomson, *System of Chemistry*, 3rd ed. (London, 1807); Dalton, *A New System of Chemical Philosophy* (London, 1808).

88. "the general conclusion." "On the Compound Sulphuret," *Retrospect*, 168–169.

89. "confirmation of Dalton's theory." Thomas Thomson, "On Oxalic Acid," *Philosophical Transactions* 98 (1808): 63–95; Wollaston, "On Super-Acid and Sub-Acid Salts," 96–102; Usselman, "Multiple Combining Proportions," 257; *Foundations of the Atomic Theory, Alembic Club Reprints, No. 2* (Edinburgh/Chicago, 1911).

89. "about theoretical chemistry." Greenaway, *John Dalton and the Atom*, 150–152.

89. "the combining proportions of gases." John Dalton, "Experimental Enquiry into the Proportions of the Several Gases or Elastic Fluids Constituting the Atmosphere," *Manchester Memoirs* (2nd series) 1 (1805): 244–258.

89. "'Dalton, Gay Lussac, Smithson, and Wollaston.'" "Review of *Elements of Chemical Philosophy*, by Sir Humphry Davy, London, 1812," *The Quarterly Review* 8 (1812): 77.

89. "as part of his atomic theory." Robert Siegfried, "Further Daltonian Doubts," *Isis* 54 (1963): 480–481; Young, "Review of Proust," 425.

90. "when 'properly considered.'" James Smithson, "On a Saline Substance from Mount Vesuvius," *Philosophical Transactions* CIII (1813): 256. "Smithson to Berzelius," 14 Nov. 1818, Royal Swedish Academy of Sciences Archives; a copy of this letter is in the Smithsonian Institution Archives, RU 7000.

7. ON THE COMPOSITION OF ZEOLITE

91. "species that he named 'zeolites,'" Axel Fredric Cronstedt, *An Essay towards a System of Mineralogy* (London, 1770), 116; Cronstedt first described this phenomenon in 1756 in the *Transactions* of the Royal Swedish Academy of Sciences; it has long been assumed that the material Cronstedt studied is the modern "stilbite," but C. Colella and A. F. Gualtieri question this in "Cronstedt's zeolite," *Microporous and Mesoporous Materials* 105 (2007): 213–221. The "frothy" appearance of these crystals when heated is now understood to be caused by the conversion of water in the crystal's cavities into steam.

92. "he met Hutton on that same trip." Faujas de Saint-Fond, *Minéralogie des volcans* (Paris, 1784), 452–454. My thanks to Jeff Gorman for this translation. James Hutton, *Theory of the Earth, with Proofs and Illustrations* (Edinburgh, 1795), 1:155–156n. This distinction is now thought to be erroneous. Dennis R. Dean, *James Hutton and the History of Geology* (Ithaca, NY: Cornell University Press, 1992), 14. Faujas and Hutton were both vulcanists, so it is perhaps not surprising that some neptunists took exactly the opposite position, arguing that the presence of zeolite crystals proved that basalt was not a product of heat at all, but had been formed by crystallization out of solution. John Pinkerton, *Modern Geography* (1804), 1:181, refers to Richard Kirwan (a neptunist) as making this very point.

92. "not considered sufficient to define them." Richard Kirwan, *Elements of Mineralogy*, 3rd ed. (London, 1810), 1:246; Thomas Thomson, *A System of Chemistry* (London, 1802), 3:433, 481.

93. "members of the new zeolite 'family.'" Haüy's first publication on zeolites was in "Observations sur les zéolithes," *Journal des Mines* 44 (1796): 86–88, but his most complete statement of the distinction between mesotype and natrolite (and what Smithson responded to) is found in M. l'Abbé Haüy, *Tableau comparatif des résultats de la cristallographie et de l'analyse chimique* (Paris, 1809), 194, 227–228.

93. "natrolite's chemical composition." M. L. P. Dejussieu, "On the Union of Natrolite with Mesotype," *Journal des Mines* 31 (1812): 201–206. Here Dejussieu describes an optical technique that Haüy used to compare tiny crystals informally.

93. "On the Composition of Zeolite." Indeed, Klaproth had named the mineral "natrolite" in reference to the fact that it contained sodium; M. H. Klapoth, "Chemische Untersuchung des Natroliths," *Gesellschaft Naturforschender Freunde zu Berlin Neue Schriften* 4 (1803): 243–248; Louis Vauquelin, "Analyse de la zéolithe de Ferroé," *Journal des Mines* 46 (1798): 576. Vauquelin analyzed "a needle zeolite from Ferro." James Smithson, "On the Composition of Zeolite," *Philosophical Transactions* 101 (1811): 171–177.

93. "why Haüy was unaware of it." Smithson, "On the Composition of Zeolite," 172. For information about Black's use of Hutton's discovery, see Thomson, *A System of Chemistry*, 481. Robert Kennedy, "Chemical Analysis of an Uncommon Species of Zeolite," *Transactions of the Royal Society of Edinburgh* 5, Part II (1802). The "cubical zeolites" that Smithson collected on Staffa are now called chabazite.

93. "'that species which Mr. Haüy calls mesotype.'" Vauquelin, "Analyse de la zéolithe de Fer-roé," 576; Smithson, "Zeolite," 172.

94. "'did or did not differ in their composition.'" Haüy was describing pyramid-shaped crystals with variations that departed from the unit form. Smithson, "Zeolite," 172.

94. "variants of the same mineral species." For a direct comparison of Smithson's analysis of mesotype with Klaproth's analysis of natrolite, see J. A. H. Lucas, *Tableau méthodique des espèces minerals* (Paris, 1813), 2:230.

94. "consequences for Smithson a few years later." Smithson, "Zeolite," 101, 175.

94. "'between these two minerals.'" "On the Composition of Zeolite.—By James Smithson," *Retrospect of Philosophical, Mechanical, Chemical, and Agricultural Discoveries* (1812), 7:166; James Sowerby, *Exotic Mineralogy* (London, 1817), 2:57.

94. "a practice also followed by others." Thomas Thomson, *A System of Chemistry*, 5th ed. (London, 1817), 3:315, 318, 320. Thomson's "zeolite family" corresponds closely to the modern use of the term "zeolite."

95. "'more and more approaches perfection.'" Dejussieu, "On the Union of Natrolite with Mesotype," 201–206.

95. "the analysis of 'Herrn Smithson.'" This article is discussed in Thomas Thomson, "Chemical Analysis of the Needlestone from Kilpatrick, in Dumbartonshire," *Annals of Philosophy* 16 (1820): 403; J. N. Fuchs, "Über die Zeolithe," *Journal für Chemie und Physik* 18 (1816): 3. Vauquelin also acknowledged his error. James Smithson, "Mémoire sur la composition de la zéolite," *Journal de Physique* 79 (1814): 145.

95. "a class of matter called 'earths.'" *A Journal of Natural Philosophy, Chemistry and the Arts* 30 (1812): 133–137; *The Philosophical Magazine* 38 (1811): 30–34; *Journal de Physique* 79 (1814): 144–149; *Annalen der Physik* (1813), 13:240–246; *American Mineralogical Journal* 1 (1814): 182–187.

95. "earths as a group of discrete materials." In 1756 Joseph Black wrote that "there plainly appeared to me to be very different kinds [of earths], altho' commonly confounded together under one name." From Joseph Black, "Experiments upon Magnesia Alba, Quicklime, and some other Alcaline Substances," *Essays and Observations, Physical and Literary. Read before a Society in Edinburgh, and Published by Them* (Edinburgh, 1756 [1898]), 2:7.

95. "magnesium, aluminum, and silicon." Torbern Bergman, *Outlines of Mineralogy*, 1783, 38, #86; Ursula Klein and Wolfgang Lefèvre, *Materials in Eighteenth-Century Science: A Historical Ontology* (Cambridge, MA: MIT Press, 2007), 169–171.

95. "and in his work on zeolites." *Annals of Philosophy* 1 (1813): 467; *Annals of Philosophy* 2 (1813): 238.

96. "'any other acknowledged mineral acid, in this zeolite.'" Cronstedt, *An Essay towards a System of Mineralogy* (1788), 1:121. This comment comes from a footnote by the editor, Gustav von Engestrom. Smithson's copy of this book is now in the Smithsonian's Cullman Library. Smithson, "Zeolite," 176.

96. "willing to concede this point." Charles Daubeny, *An Introduction to the Atomic Theory* (Oxford, 1831), 93.

97. "'silicates, either simple or compound.'" Smithson, "Zeolite," 176.

97. "the Swedish chemist Jacob Berzelius." This quote was the first time that the term "silicates" appeared in the *Philosophical Transactions*, and it has been suggested that Smithson coined it. See, for example, *The Quarterly Journal of the Geological Society of London* 5 (1849): xxiv. However, "silicates" had been in use for some time as a name for the "oxides" of silica—"silicates" in this context referring essentially to materials containing silica. By contrast, silica acting as an acid allowed Smithson to use the term "silicates" to refer to chemically neutral salts, which in turn allowed them to be distinguished into separate species.

97. "could be explained by this assumption." J. Berzelius, *Attempt to Establish a Pure Scientific System of Mineralogy, by the Application of the Electro-Chemical Theory and the Chemical Proportions* (London, 1814), 12, 27–42.

98. "'was reserved for Mr. Smithson.'" Thomas Thomson, "Berzelius's System of Mineralogy," *Annals of Philosophy* 5 (1815): 304–305; Thomas Thomson, *A System of Chemistry* (Edinburgh, 1817), 1:251; Thomas Thomson, *The History of Chemistry* (London, 1831), 2:46.

98. "'ingénieuse observation de M. Smithson.'" Humphry Davy, "On a Deposit Found in the Waters of the Baths of Lucca," *The Collected Works of Sir Humphry Davy, Vol. VI*, ed. John Davy (London, 1840), 205. Note that this article was published in *Mémoires de l'Académie de Naples* and the *Annales de Chimie et de Physique* (1821): 194–196; Humphry Davy, "Letter 107, Humphry Davy to Faraday, 16 November 1819," *The Correspondence of Michael Faraday, Vol. 1: 1811–1831*, ed. Frank A. J. L. James (Stevenage, UK: Institution of Engineering and Technology, 1991), 186. See, for example, Humphry Davy, "Mémoire sur un dépôt trouvé dans les eaux de Lucca," *Annales de Chimie et de Physique* (1821): 195; Andrew Ure, *Dictionnaire de chimie* (Paris, 1824), 4:425.

98. "oxides should have the same property." William Brande, "Review of a Work entitled 'An Attempt to Establish a Pure Scientific System of Mineralogy, by the Application of the Electro-chemical Theory and the Chemical Proportions,' By J. Jacob Berzelius," *Quarterly Journal of Science, Literature, and the Arts* 1 (1816): 226. Specifically, Brande wrote: "As to [silica] performing the functions of an acid (if we understand the term) so do phosphorus, sulphur, etc."; William Brande, "Art. XV. A System of Chemistry, in four volumes, 8 vo. by Thomas Thomson [Review]," *Quarterly Journal of Science, Literature, and the Arts* 4 (1818): 318.

98. "'any well informed chemist.'" John Murray, *A System of Chemistry*, 4th ed. (Edinburgh, 1819), 3:97–98; Franklin Bache, *A System of Chemistry for the Use of Students of Medicine* (Philadelphia, 1819), 66. A similar objection appears in R. Phillips, "Observations on Certain Substances Which Have Been Supposed to Act as Acids, and as Alkalies," *Annals of Philosophy* 4 (1822): 53–54. Thomas Thomson, "Answer to the Review of the Sixth Edition of Dr. Thomson's System of Chemistry," *Annals of Philosophy* 3 (1822): 248–249.

99. "how closely his articles were read." See Wladislaw Asch and Dagobert Asch, *The Silicates in Chemistry and Commerce* (London: Constable, 1914), 3; Abraham Rees, *The Cyclopaedia; or Universal Dictionary* (London: Longman, 1819), 32:12.

99. "that Smithson had left his fortune to establish." C. M. Élie de Beaumont, "Memoir of Oersted," read at the annual public sitting of the Academy, December 29, 1862; C. M. Élie de Beaumont, "Memoir of Oersted," *Annual Report of The Board of Regents of the Smithsonian Institution* (Washington, DC, 1869), 173.

8. ULMIN

---

101. "one related to the processes of life." James Smithson, "On a Substance from the Elm Tree called Ulmin," *Philosophical Transactions* 102 (1813): 64–70.

102. "he published his analysis." Vauquelin also worked in partnership with the influential French chemist Antoine Fourcroy. See W. A. Smeaton, *Fourcroy, Chemist and Revolutionary, 1755–1809* (Cambridge: Heffer and Sons, 1962), 166–176; Citoyen Vauquelin, "Observations sur une maladie des arbres, et spécialement de l'orme (ulmus campestris Lin.), analogue à un ulcère," *Mémoires de l'Institut National des Sciences et Arts* 2 (1797): 23–30.

102. "fell roughly into two groups." Turnsole's natural color is blue, but when exposed to an acid it turns red and when exposed to an alkali it turns green. Using turnsole that has already been exposed to an acid (and is thus red) was a technique chemists used to increase its sensitivity to alkalis. Thus Vauquelin's report that the solution he tested turned red turnsole green, confirmed that it was alkaline. It is likely that the two substances Vauquelin found issuing from the wound were identical and that one was simply darkened from exposure to light and air. He also described a third "lightly colored," material around the wound that did not dissolve in water, but he failed to provide any further description.

103. "they changed it into a thick paste." For more on the medicinal use of metallic oxides, see Erasmus Darwin, *Zoonomia; or, The Laws of Organic Life*, 3rd ed. (London, 1801), 2:503–505.

103. "the German chemist Martin Klaproth." An entry in Smithson's notebook states that a gumlike substance had been sent him "adhering to the bark from Palermo in Sicily by Dr Thomson with the following label 'Saline gum from an old elm tree under my window. Palermo June 1800.'" from "Smithson's Mineral Notes," Smithsonian Institution Archives (SIA), RU 7000, Box 2.

103. "'any detailed observation reported about it.'" Martin Klaproth, "Chemische Untersuchung eines gummigen Pflanzensaftes von Stamm eines Ulme," *Journal der Chemie* (1805): 329–331. The report was read to the Academy on July 28, 1802. My thanks to Jeff Gorman for translating this article.

104. "'ulmin,' from the Latin *ulmis* (elm)." Thomas Thomson, *A System of Chemistry*, 3rd ed. (Edinburgh, 1807); Thomas Thomson, *A System of Chemistry* (Edinburgh, 1810), 4:695–696.

104. "'in general they dissolve in alcohol.'" Ibid., 637–638. I have added the title for the first category, which is missing in Thomson's 1810 list. He used this term in the 1817 edition of his book. Smithson cites the 1810 edition in his article on Ulmin.

105. "'the substance which I examined.'" Thomas Thomson, "On Ulmin," *Annals of Philosophy* 1 (1813): 23–26.

108. "published in the *Journal de Physique*." Thomas Thomson, "Philosophical Transactions for 1813, Part I," *Annals of Philosophy* 2 (1814): 229–230; "Proceedings of Learned Societies. Royal Society," *The Philosophical Magazine* 40 (1812): 454; James Smithson, Esq. F.R.S., "On a Substance from the Elm Tree, called Ulmin," *The Philosophical Magazine* 42 (1813): 204–208. "[Review] On a Substance from the Elm Tree, called Ulmin. By James Smithson," *The Monthly Review* 74 (1814): 73; "On a Substance from the Elm-Tree called Ulmine. By James Smithson," *Retrospect of Philosophical, Mechanical, Chemical and Agricultural Discoveries* 8 (1815): 365–356; M. Smithson, "Expériences sur l'ulmine," *Journal de Physique* 78 (1814): 311–315; Smithson submitted a two-page long list of errata that appeared in the following issue (150–151). Smithson's analysis of ulmin was noted in the issue after that in a discussion of "la chimie des végétaux," *Journal de Physique* 80 (1815): 76.

108. "'Mr. Smithson, well known for his precision.'" Thomas Thomson, "On Ulmin," *Annals of Philosophy* 1 (1813): 73; Thomas Thomson, *A System of Chemistry*, 5th ed. (Edinburgh, 1817), 4:48.

108. "'either by Mr. Smithson or myself.'" Thomas Thomson, "Article XIV. Scientific Intelligence; and Notices of Subjects connected with Science. 1. Ulmin," *Annals of Philosophy* 2 (1814): 395–396.

108. "in the usual manner with hot water." Berzelius communicated this to Thomson in a letter: "V. Ulmin," *Annals of Philosophy* 2 (1814): 314; Thomson, *A System of Chemistry* (5th ed.) 4:48–49.

109. "reprinted in *The Repertory*." Henri Braconnot, "Sur la conversion du corps ligneux en gomme, en sucre, et en un acide d'une nature particulièr, par le moyen de l'acide sulfurique," *Annales de Chimie et de Physique* 12 (1819): 172–195; Smithson is mentioned on page 190. An English translation of Braconnot's article appeared in *The Repertory of Arts, Manufactures, and Agriculture* 36 (1820): 246–255.

109. "the coal used in its manufacture." *Journal de Pharmacie et des Sciences Accessoires* 6 (1820); Wilfrid Francis and H. M. Morris, "Relationship Between Oxidizability and Composition of Coal," *U.S. Department of Commerce Bulletin* 340 (1931): 1; Samuel L. Dana, "Appendix to Chapter IV. History of Geine," *A Muck Manual for Farmers* (Lowell, MA: Daniel Bixby, 1842), 74.

109. "it was frequently acknowledged." "Chemical Science," *Journal of the Royal Institution of Great Britain* 1 (1831): 179.

109. "a good chance of being called ulmin." See, for example, M. Polydore Boullay, "Dissertation sur l'ulmine," *Annales de Chimie* 43 (1830): 273–285; William Thomas Brande, *A Manual of Chemistry* (London, 1819), 365–366. Brande was professor of chemistry at the Royal Institution.

109. "'rejection of the word ulmin.'" Jöns Jakob Berzelius, *Traité de chimie* (*Chimie organique*) (Paris, 1832), 6:240. My thanks to Jeff Gorman for this translation.

110. "'or solely in mere nomenclature.'" Rev. John M. Wilson, *The Rural Cyclopedia, or a General Dictionary of Agriculture* (Edinburgh, 1848), 2:759.

110. "'a difference about names, not things.'" Dana, A *Muck Manual for Farmers*, 72.

9. A SALINE SUBSTANCE FROM MOUNT VESUVIUS

111. "an explanation for volcanoes and their products." James Smithson, "On a Saline Substance from Mount Vesuvius," *Philosophical Transactions* 103 (1813): 256–262. The paper was read to the Royal Society on July 8, 1813. The use of the term "saline substance" refers to the chemical combination of an alkali (earth, or metallic oxide) with an acid, to produce a "salt."

112. "'would probably add to their number.'" Among James Smithson's surviving papers is a packing list, in William Thomson's handwriting, dated November 22, 1796, and describing volcanic materials that Thomson had sent him. Item 7 is "vitrolated tartar" from the cone of Vesuvius, as described in Smithson's "Saline Substance" paper. Smithsonian Institution Archives, RU7000, box 2, f.2.

112. "to discover seven new chemical elements." Humphry Davy, "The Bakerian Lecture, on Some New Phenomena of Chemical Changes Produced by Electricity, Particularly the Decomposition of the Fixed Alkalies, and the Exhibition of the New Substances Which Constitute Their Bases; and on the General Nature of Alkaline Bodies," *Philosophical Transactions* 98 (1808): 2.

113. "into his small samples—to great effect." Volts are not a measure of current, so it is more precise to say that Davy's batteries produced a direct current with a *potential* of 250 volts. The voltage calculation for the three voltaic piles Davy used is based on the total number of plates and the conservative assumption that each set of plates produced at least one volt when placed in series. For Davy's description of his batteries, see ibid., 3, 6. See also Alonzo Gray, *Elements of Chemistry; Containing the Principles of the Science, Both Experimental and Practical* (New York, 1841), 81–82, and Humphry Davy, *Elements of Chemical Philosophy* (London, 1812), 80–81.

113. "'they were formed, and others remained.'" Potassium hydroxide (KOH) is a form of lye, and is a strong base. See Humphry Davy, "On Some Chemical Agencies of Electricity," *Philosophical Transactions* 96 (1806): 5.

113. "but no longer." Ibid., 27.

113. "this new metal was extremely reactive." This was how Davy described the potassium he produced. We now know that pure potassium is solid up to 70°C, so it is likely that his potassium was impure. Based on its reported physical properties, it may have been alloyed with as much as 10 percent metallic sodium, an amount that Davy would have had a hard time detecting.

114. "have to do with volcanoes and geology?" For the years 1806–1810, Davy's five published Bakerian lectures are found, respectively, in volumes 97 through 101 of the *Philosophical Transactions*.

114. "the scientific understanding of the Earth." Thomas Edward Thorpe, *Humphry Davy: Poet and Philosopher*, ed. Sir Henry E. Roscoe (London: Century Science Series, 2017), 23.

114. "the result of a violent chemical reaction." Henry Cavendish, "Experiments to Determine the Density of the Earth," *Philosophical Transactions* 88 (1798): 469–526. Cavendish's extremely accurate methods found that Earth's mean density was 5.48 times the density of water—within one percent of the modern value! This investigation followed up on the 1775 report of Nevil Maskelyne, who (Cavendish reported) measured Earth's mean density as 4.5 times the density of water. This work had earned Maskelyne the Royal Society's Copley medal: Nevil Maskelyne, "An Account of Observations Made on the Mountain Schehallien for Finding Its Attraction," *Philosophical Transactions* 65 (1775): 500–542. In 1778 Charles Hutton used Maskelyne's data to suggest that "nearly 2/3 of the diameter of the earth is the central or metalline part," Charles Hutton, *An Account of the Calculations Made from the Survey and Measures Taken at Schehallien in Order to Ascertain the Mean Density of the Earth* (London, 1778), 96. In 1811 Hutton revisited this topic and, using Cavendish's improved data, reached the same conclusion. Charles Hutton, "Letter from Dr. Hutton on the Calculations for Ascertaining the Mean Density of the Earth," *Philosophical Magazine* 38 (1811): 112–116.

115. "but his wealth and taste as well." e.g., William Hamilton, *Observations on Mount Vesuvius, Mount Etna, and Other Volcanos: In a Series of Letters Addressed to the Royal Society* (London, 1774).

115. "more than fifty hand-colored prints." Sir William Hamilton, *Campi Phlegraei* (London, 1776).

115. "the perfect illusion of a current of glowing lava." William Hamilton, "Letter of Dec. 29, 1767, 'Hamilton to the Earl of Morton, President of the Royal Society,'" *Observations on Mount Vesuvius, Mount Etna, and Other Volcanos*; Maiken Umbach, *Federalism & Enlightenment in Germany, 1740-1806* (London: Hambledon Press, 2000), 68; Bent Sørensen, "Sir William Hamilton's Vesuvian Apparatus," *Apollo* 159 (2004): 50–57.

116. "burning sulfur added to heighten the experience." Nicholas Daly, *The Demographic Imagination and the Nineteenth-Century City* (London: Cambridge University Press, 2015), 22–25. Covent Garden advertisement from *The Morning Chronicle and London Advertiser*, 1780; Richard D. Altick, *The Shows of London* (Cambridge, MA: Harvard University Press, 1978), 96.

116. "mechanism by which real volcanoes were formed." For example, Humphry Davy, "On Some New Phenomena of Chemical Changes," *Philosophical Transactions* 98 (1808): 44.

116. "'cheering of the audience [filled the auditorium].'" Davy's pyrite models appear to have been based on models that Martin Lister made in the seventeenth century. Simon Werrett,

Fireworks, Pyrotechnic Arts & Sciences in European History (Chicago: University of Chicago Press, 2010), 217. The quotation is from John Ayrton Paris, *The Life of Sir Humphry Davy* (London, 1831), 1:306.

117. "'stony matter analogous to lavas.'" Humphry Davy, "Electro-Chemical Researches, on the Decomposition of the Earths," *Philosophical Transactions* 98 (1808): 369. In 1828 Davy published a more detailed description of his theory, which by then he had abandoned: Humphry Davy, "On the Phenomena of Volcanoes," *Philosophical Transactions* 118 (1828): 241–250.

117. "'relate to the future order of things.'" Report on Davy's geology lecture in *The Bath Chronicle*, April 25, 1811. This article also appeared in *The Lancaster Gazette and General Advertiser*, May 18, 1811. Alexander Tilloch, "Mr. Davy's Lectures on Geology—No. 1," *Philosophical Magazine* 37 (1811): 392–393, 468–470. Faraday appears to confirm this in his 1811 notebook in the Royal Institution Archives, where he describes the model volcano as being "illustrative of what he [Davy] had advanced."

117. "widely seen as models of these values." Quote from Basil Willey, *The Eighteenth Century Background* (New York: Columbia University Press, 1941), 136. Historians now reject the "conflict theory," which proposes an inherent conflict between science and religion. See, for example, Lawrence Principe, *Science and Religion* (Chantilly, VA: The Teaching Company, 2006); Helena Rosenblatt, "The Christian Enlightenment," *The Cambridge History of Christianity*, ed. Stewart J. Brown and Timothy Tackett (Cambridge: Cambridge University Press, 2006), 7:284; Clarissa Campbell Orr, "Queen Charlotte, 'Scientific Queen,'" *Queenship in Britain 1660–1837, Royal Patronage, Court Culture and Dynastic Politics* (Manchester: Manchester University Press, 2002), 249.

118. "'without admiring the Creator? I think not!'" Paul A. Tunbridge, "Jean André De Luc, F. R. S. (1727–1817)," *Notes and Records of the Royal Society of London* 26 (1971): 18. My thanks to Jeff Gorman for this translation.

118. "to have free access to him." Clarissa Campbell Orr, "Queen Charlotte as Patron: Some Intellectual and Social Contexts," *The Court Historian* 6 (2001): 199–200; Tunbridge, "Jean André De Luc," 15–33.

118. "signaled her approval of it." J. A. de Luc, *Lettres physiques et morales sur l'Histoire de la Terre et de l'Homme. Adressées à la Reine de la Grande Bretagne* (Paris, 1779); *Letters Philosophical and Moral Concerning the History of the Earth and of Man; Addressed to the Queen of Great Britain*, 5 vols. (The Hague, 1780).

119. "that the deists and atheists claimed it to be." For an overview of the six periods, see Gordon L. Davies, *The Earth in Decay: A History of British Geomorphology 1578–1878* (New York: Elsevier, 1969), 135–136; Roy Porter, *The Making of Geology: Earth Science in Britain 1660–1815* (Cambridge: Cambridge University Press, 1977), 199; De Luc, *Lettres physiques et morales* (Paris, 1779), discourse 2, 1:23–52, quotation at 24; translation from Martin Rudwick, *Bursting the Limits of Time: The Reconstruction of Geohistory in the Age of Revolution* (Chicago: University of Chicago Press, 2005), 153.

119. "'philosophy against Deism and atheism.'" Rudwick, *Bursting*, 155–157. "Art. VI, Lettres Physiques et Morales," *The Monthly Review or Literary Journal* 62 (1780): 527–528; Cherry L. E. Lewis, "'Our favourite science': Lord Bute and James Parkinson searching for a Theory of the Earth," *Geology and Religion* (2009): 116; John William Fletcher, "Eulogy on the Christian Philosophers," *The Works of the Reverend John Fletcher, Late Vicar of Madeley* (London, 1860), 9:393.

119. "suspicious of new scientific ideas." Samuel Bernstein, "English Reactions to the French Revolution," *Science & Society* 9 (1945): 147–171; Norton Garfinkle, "Science and Religion in England, 1790–1800," *Journal of the History of Ideas* 16 (1955): 376–388; Edward Alexander, *Matthew Arnold and John Stuart Mill* (Malden, MA: Routledge, 2010), 209.

120. "'addicted to [radical] Politicks.'" S. T. Coleridge, review of Beddoes,' "A Letter to the Right Hon. William Pitt, in The Watchman (1796)," *The Collected Works of Samuel Taylor Coleridge: The Watchman*, ed. L. Patton and K. Paul (Malden, MA: Routledge, 1970), 100; Yale University, Beinecke Rare Book and Manuscript Library, Osborn Shelves, f.d, 10/23, Rumford (April 1804), quoted in John Gascoigne, *Joseph Banks and the English Enlightenment* (Cambridge: Cambridge University Press, 2003), 252–253.

120. "geological theory from it—which he did." Porter, *The Making of Geology*, 198. At least some of this distrust of scientific speculation, or "theory," grew out of a broader English rejection of the "empire of reason" that took place in the second half of the eighteenth century. Simpson has written about the nationalistic aspect of this impulse, which was popularly understood as "British common sense" and used to undermine French and German influences. David Simpson, *Romanticism, Nationalism, and the Revolt against Theory* (Chicago: University of Chicago Press, 1993). J. Rennell to John Hunter (n.d.), transcript in the Royal College of Surgeons, ms. 49 c 2, 102–105; Rudwick, *Bursting the Limits of Time*, 126–127; Royal Botanic Gardens Library, Kew, Banks Correspondence, quoted in Gascoigne, *Joseph Banks and the English Enlightenment*, 114.

120. "little of that discussion appeared in print." Gordon L. Herries Davies, *Whatever Is under the Earth: The Geological Society of London, 1807–2007* (London: Geological Society of London, 2007), 30, 40.

121. "and throughout Smithson's lifetime." Faujas de St. Fond, *Essai de géologie* (Paris, 1803), 1:19–20; David Oldroyd, *Thinking about the Earth: A History of Ideas in Geology* (Cambridge, MA: Harvard University Press, 1996), 146; Rudwick, *Bursting the Limits of Time*, 431.

121. "a firm supporter of the reality of the Deluge." Georges Cuvier, *Essay on the Theory of the Earth* (Edinburgh, 1813); Oldroyd, *Thinking about the Earth*, 133, 146. G. B. Greenough, *A Critical Examination of the First Principles of Geology; in a Series of Essays* (London, 1819); W. D. Conybeare and Will Phillips, *Outlines of the Geology of England and Wales, with an Introductory Compendium of the General Principles of That Science* (London, 1822); Norman Cohn, *Noah's Flood: The Genesis Story in Western Thought* (New Haven, CT: Yale University Press, 1996), 116; William Buckland, *Vindiciae Geologicae; or the Connexion of Geology with Religion Explained* (Oxford: Oxford University Press, 1820).

121. "the other was the article by Smithson." S. Tillard, "A Narrative of the Eruption of a Volcano in the Sea off the Island of St. Michael," *Philosophical Transactions* 102 (1812): 152–158.

123. "only Davy's theory could explain." Smithson, "On a Saline Substance from Mount Vesuvius," *Philosophical Transactions* 103 (1813): 261.

123. "'Mr. Smithson is a decided Huttonian.'" "Review. On a Saline Substance from Mount Vesuvius. By James Smithson, Esq., F.R.S.," *The Monthly Review* 74 (1814): 162; "[Review] On a Saline Substance, from Mount Vesuvius. By James Smithson," *The Eclectic Review* 4 (1815): 70–71; "[Review] An Analysis of a Substance Thrown out of Mount Vesuvius, by James Smithson," *The Medical and Physical Journal* 30 (1814): 424–425; *Annals of Philosophy* 3 (1814): 23–24; Buffon, *Histoire naturelle*, vol. 1; Sara Schechner, *Comets, Popular Culture, and the Birth of Cosmology* (Princeton: Princeton University Press, 1997), 198–200, 208.

123. "determined to refute it in any way he could." J. A. de Luc, "Remarques sur la théorie géologique avancée par James Smithson," *Journal de Physique* 78 (1813): 386–398; J. A. de Luc, "Remarks on the Geological Theory Supported by James Smithson," *Philosophical Magazine* 43 (1813): 127–137. We know that he also submitted this article to Nicholson's *Journal*, which declined to publish it. Ibid., 137.

123. "this was all Smithson's idea." The question of whether de Luc was actually aware of Davy's theory is a fair one, since no direct evidence of this has yet been found. But from de Luc's letters it is clear that he was actively reading both the popular and scientific journals for articles critical of his theory. See, for example, "De Luc to Nares, July 8, 1794," in Edward Nares and George Cecil White, *A Versatile Professor* (London: Johnson, 1903), 280–282. One of the many ways that de Luc could have become aware of Davy's ideas was from Alexander Tilloch, editor of the *Philosophical Magazine*, who apparently attended Davy's geology lectures at the Royal Institution, and in 1811 published this report: "Mr. Davy remarked, that the source of this imaginary fire [the plutonist's 'central fire'] might be attributed to the existence of the earths in their metallic state in the interior, acted on by air and water, and thus supplying [the volcano's] fuel." Alexander Tilloch, "Mr. Davy's Lectures on Geology—No. 1," *Philosophical Magazine* 37 (1811): 392–393. This is a journal to which de Luc frequently submitted articles. His own article criticizing Smithson would be published in this journal in 1814.

124. "'in which I shall now follow him.'" De Luc, "Remarks on the Geological Theory Supported by James Smithson," 127–128.

124. "several such crystals in his personal collection." Smithson, "On a Saline Substance," 257. De Luc reported that his crystals had been found in a layer of primitive rock in the central ridge of the Alps, an area that he himself maintained had once been at the bottom of the ocean. He failed to discuss what would be found in primitive strata that had never been submerged. De Luc, "Remarks," 130–135.

125. "it was as if the article had never been written." An example of how de Luc's criticisms were understood by nonspecialists can be seen in Rev. James Little, *Conjectures on the Physical Causes of Earthquakes and Volcanoes* (Dublin, 1820), 79–80, 83. The confusion of authorship came from S. Breislak, *Institutions géologiques* (Milan, 1818), 135. See also J. F. Krüger, *Geschichte der Urwelt* (Leipzig, 1822), 160; *Journal für Chemie und Physik* 15 (1815): 243.

125. "served in Banks's place if he was unavailable." Royal Society minutes, July 8, 1813, JB041, Royal Society Archives. Minutes of the committee on Papers, Royal Society archives, July 15, 1813, 254–256.

126. "or he decided to leave on his own, we do not know." Banks had agreed to give special treatment to one of de Luc's articles. J. A. de Luc, "Communications on the Mode of Action of the Galvanic Pile announced," *Journal of Natural Philosophy* 26 (1810): 69–70.

126. "'has been extinguished only at its surface.'" Sir Alexander Crichton, "On the Climate of the Antediluvian World, and Its Independence of Solar Influence; and on the Formation of Granite," *Annals of Philosophy* 9 (1825): 107–108; M. L. Cordier, "On the Temperature of the Interior of the Earth," *The Edinburgh New Philosophical Journal* 5 (1828): 278.

126. "and it faded into obscurity." Humphry Davy, "On the Phenomena of Volcanoes," *Philosophical Transactions* 118 (1828): 241–250. This article concludes: "the hypothesis of the nucleus of the globe being composed of fluid matter, offers a still more simple solution of the phaenomena of volcanic fires than that which has been just developed." Davy was in poor health when he wrote this and was likely putting his affairs in order.

10. THE COLOURING MATTERS OF SOME VEGETABLES

127. "the Colouring Matters of Some Vegetables." James Smithson, "A Few Facts Relative to the Colouring Matters of Some Vegetables," *Philosophical Transactions* 108 (1818): 110–117.

128. "whether a substance was acid or alkali." Robert Boyle, *Experiments and Considerations Touching Colours* (London, 1664), 245–246; Allen G. Debus, *The Chemical Philosophy: Paracelsian Science and Medicine in the Sixteenth and Seventeenth Centuries* (New York: Dover, 2002), 491; Robert Boyle, "Reflections upon the Hypothesis of Alkali and Acidum," *The Works of the Honourable Robert Boyle, Esq., epitomiz'd* (London, 1700), 4:284; Marie Boas, *Robert Boyle and Seventeenth-Century Chemistry* (Cambridge: Cambridge University Press, 1958), 133, 217. A video about the preparation and use of "syrup of violet" can be seen at https://folgerpedia.folger.edu/Beyond_Home_Remedy:_Women,_Medicine,_and_Science#Recipe:_To_Make_Sirrop_of_Violets.

129. "by which they worked was a mystery." Frederick Accum, *A Practical Essay on Chemical Re-Agents, or Tests* (Philadelphia, 1817). Accum described using litmus, litmus paper, red cabbage, Brazilwood, and turmeric. James Louis Macie, "An Account of Some Chemical Experiments on Tabasheer," *Philosophical Transactions* 81 (1791): 371; Torbern Bergman, *Physical and Chemical Essays*, trans. Edmund Cullen (London, 1788).

129. "does not seem to have been aware of this." For an extensive bibliography on turnsole, see William Eamon, "New Light on Robert Boyle and the Discovery of Colour Indicators," *Ambix* 27 (1980): 205; "Method of Preparing the Dutch Turnsol Blue," *The Philosophical Magazine* 4 (1799): 17–18.

130. "It was an important finding." For the preparation of turnsole, see M. I. A. Chaptal, *Elements of Chemistry*, 3rd ed. (London, 1800), 1:185–187.

131. "he did not investigate it further." For an example of his use, see Bergman, *Physical and Chemical Essays*, 1:126; A. Proteaux, *Practical Guide for the Manufacture of Paper and Boards* (London: Sampson Low, Son, and Marston, 1866).

133. "just a collection of facts with no real point." See, for example, the discussion in Bernadette Bansaude-Vincent, "A View of the Chemical Revolution through Contemporary Textbooks: Lavoisier, Fourcroy and Chaptal," *British Journal for the History of Science* 23 (1990): 439–440.

133. "mentioned this in his textbook." *The Philosophical Magazine* 51 (1818): 58; "Proceedings of the Royal Society of London," *Quarterly Journal of Science, Literature, and the Arts* 4 (1818): 364. The abstract prepared for the *Philosophical Transactions* included the names Smithson suggested for his principles. "Philosophical Transactions of the Royal Society of London, for the Year 1818. Part 1," *The Monthly Review* 87 (1818): 186; "Notice des Séances de la Soc. Roy. de Londres," *Bibliothèque Universal des Sciences, Belles-Lettres, et Arts* 7 (1818): 153; William Thomas Brande, *Manual of Chemistry* (New York, 1821), 517–518.

133. "matters that he studied were anthocyanins." Mircea Enachescu Dauthy, "2.2 Chemical Composition," *Fruit and Vegetable Processing*, FAO Agricultural Services Bulletin No. 119 (Rome: FAO, 1995).

133. "improvements in purification methods." Kumi Yoshida, Mihoko Mori, and Tadao Kondo, "Blue Flower Color Development by Anthocyanins: From Chemical Structure to Cell Physiology," *Natural Product Reports* 26 (2009): 884–915.

134. "'Chemistry of Anthocyanins' (1916)." Ibid., 885; Richard Willstätter and A. E. Everest, "Ueber den Farbstoff der Kornblume," *Justus Liebigs Annalen der Chemie* 401 (1913): 189–232.

The bibliography "Chemistry of Anthocyanins" is in Muriel Wheldale Onslow, *The Anthocyanin Pigments of Plants* (Cambridge: Cambridge University Press, 1916), 241.

134. "'made his bequest to the United States.'" In 1880, on the occasion of the fiftieth anniversary of Smithson's death, the Smithsonian Institution made a public request for information about its founder. Among the long list of specific questions was the following: "What can be learned of the disagreement between Mr. Smithson and the Council of the Royal Society?" This notice appeared in *Nature* 22 (1880): 110, and was copied in *The New York Times*, June 27, 1880. In *The Lost World of James Smithson* (New York: Bloomsbury, 2007), Heather Ewing reports that "Joseph Henry related Wheatstone's story in a letter to General John Henry Lefroy, February 12, 1878. Henry had spent a day with Wheatstone in London in 1870; Henry to Caroline Henry, July 1, 1870; Joseph Henry Papers." This is the source of the account in William J. Rhees, "James Smithson and His Bequest," *Smithsonian Miscellaneous Collections* 330 (1880): 21–22. "The United States National Museum," *Bulletin of the International Bureau of the American Republics* 29 (1909): 869; Louis Agassiz, "Letter on the Smithsonian Institution," *American Journal of Science* 19 (March 1855): 284–287.

11. A SULPHURET OF LEAD AND ARSENIC, AND "PLOMB GOMME"

136. "as much as he could say." James Smithson, "On a Native Compound of Sulphuret of Lead and Arsenic," *Annals of Philosophy* 14 (1819): 96–97. Smithson probably acquired the mineral in the summer of 1805 when he was traveling in Switzerland. It is also possible, though less likely, that he got it in the summer of 1795, when he was exploring the area north of Milan. See Heather Ewing, *The Lost World of James Smithson* (New York: Bloomsbury, 2007), 156–186, 218–258.

136. "specks of metal that he now reported." R. H. Solly, "Sulpharsenites of Lead from the Binnenthal," *The Mineralogical Magazine and Journal of the Mineralogical Society* 11 (1897): 285. Lengenbach is the type locality for at least twenty-nine minerals and continues to produce new ones. See https://www.mindat.org/loc-3207.html and http://fglb.clubdesk.ch (both last accessed August 2, 2018).

136. "made many years earlier." Professor Proust, "On Metallic Sulphurets," *Philosophical Magazine*, Series 1, 21 (1805): 208; James Smithson, "On the Composition of the Compound Sulphuret from Huel Boys," *Philosophical Transactions* 98 (1808): 55.

136. "That name is still used." Thomson, *Annals of Philosophy* 16 (1820): 100; Parker Cleaveland, *An Elementary Treatise on Mineralogy and Geology*, 2nd ed. (Philadelphia, 1822), 1:660–661; *Journal de Physique* 15 (1820): 75; Solly, "Sulpharsenites of Lead from the Binnenthal," 282–290; Carl Friedrich Nauman, *Elemente der Mineralogie* (Leipzig, 1871), 543–544.

138. "'sulphuric acid, in bodies.'" William Henry, *The Elements of Experimental Chemistry*, 9th ed. (London, 1823), 2:554.

138. "sulfur would still make a brown stain." J. J. Berzelius, *Traité de chimie* (Paris, 1831), 3:342–343.

138. "detect other substances as well." J. J. Berzelius, *De l'analyse des corps inorganiques* (Paris, 1827), 31; Friedrich August Walchner, *Handbuch der gesammten Mineralogie* (Karlsruhe, 1829), 159.

138. "another one to his publisher." James Smithson, "On a Native Hydrous Aluminate of Lead or Plomb Gomme," *Annals of Philosophy* 14 (1819): 31.

139. "he had ready access to it." Romé de l'Isle, *Cristallographie* (Paris, 1783), 398–399; Gillet de Laumont, "Mémoire sur la description de plusieurs filons metallique de Bretagne," *Observations sur la Physique* 28 (1786): 385; James Sowerby, *Exotic Mineralogy: or Coloured Figures of Foreign Minerals* (London, 1817), 2:clxi. Although dated 1817, this last volume of Sowerby's works on mineralogy was completed and published in 1820.

139. "and failed to observe this property." Ewing, *Lost World*, 276; Smithson, "Plomb Gomme," 31.

139. "He was fifty-four." Sir John Barrow, *Sketches of the Royal Society* (London, 1849), 160–161.

139. "'the late Smithson Tennant.'" Smithson, "Plomb Gomme," 31; "Scientific Intelligence," *Annals of Philosophy* 13 (1819): 381.

139. "'from Huelgoat, near Poullaouen, Brittany.'" Smithson, "Plomb Gomme," 31.

140. "'Argilla [Aluminum] and Water.'" Thomas Thomson, *Annals of Philosophy* 16 (1820): 100; Sowerby, *Exotic Mineralogy*, 130–131; Cleaveland, *An Elementary Treatise on Mineralogy and Geology*, 1:634; William Phillips, *An Elementary Treatise on Mineralogy*, 3rd ed. (London, 1823), 338.

140. "it contained aluminum." Armand Dufrénoy, "Description et analyse du plomb gomme de la mine de la Nussière près Beaujeu," *Annales de Chimie* 59 (1835): 440; Jacob Berzelius, *The Use of the Blowpipe in Chemical Analysis* (London, 1822), 162.

141. "masked by the blue of the aluminum." Philip Ball, *Bright Earth* (Chicago: University of Chicago Press, 2001), 158–160; J. R. Partington, *History of Chemistry* (London: Macmillan, 1964), 4:92; Atul Singhal, *The Pearson Guide to Inorganic Chemistry for the IIT-JEE* (Delhi: Pearson, 2010), 10.10.

141. "name was no longer mentioned." Edward Dana, *Descriptive Mineralogy*, 6th ed. (New York: Wiley, 1914), 855; J. W. Mellor, *Comprehensive Treatise on Inorganic and Theoretical Chemistry* (London: Longmans, Green and Co, 1947), 7:877, 886.

141. "in Smithson's time it frequently worked." "Scientific Intelligence," *Annals of Philosophy* 2 (1814): 238.

12. FIBROUS COPPER AND CAPILLARY METALLIC TIN

143. "'crystallization was perfectly inadmissible.'" James Smithson, "On a Fibrous Metallic Copper," *Annals of Philosophy* 16 (1820): 46–47.

144. "'the object of my toil.'" Ibid., 47. Also known as "copper pyrite" in the nineteenth century. See *Chemical News* 32 (1875): 16.

145. "seems to have noticed it." "Noticias das Sciencias," *Annaes das Sciencias* 14 (1821): 64.

145. "'twisted in various directions.'" James Smithson, "On Some Capillary Metallic Tin," *Annals of Philosophy* 1 (1821): 271. Ampère's colleague may have been the chemist Nicholas Clément.

146. "'solid metal, like a small shower.'" Francis Bacon, *The Works of Francis Bacon* (London, 1889), 395.

146. "like a fine, dripping dew." "Hydrostatics," *Encyclopaedia Britannica* (Edinburgh, 1792), 9:2; John Locke, *An Essay Concerning Human Understanding* (Glasgow, 1759), 1:177.

146. "what was already known." Smithson, "Tin," 75.

146. "it was still in its collection." For example: "Chemical Science," *Quarterly Journal of Science, Literature and the Arts* 11 (1821): 385–386; "Capillarität der Metalle," *Journal für Chemie und Physik* 32 (1821): 478; *The Saturday Magazine* 10 (1821): 377; Charles Konig, "On Mr. Smithson's Hypothesis of the Formation of Capillary Copper," *Annals of Philosophy* 2 (1821): 291–292. See also James Hall, "Account of a Series of Experiments, Shewing the Effects of Compression in Modifying the Action of Heat," *Journal of Natural Philosophy, Chemistry and the Arts* 13 (1806): 338–339, and 14 (1806): 318.

146. " 'a model for scientific purposes.' " James Smithson, *The Scientific Writings of James Smithson*, ed. William J. Rhees (Washington, DC, 1879), 155.

147. "that Smithson would have appreciated." Smithson, "Tin," 271; J. W. Mellor, "Tin," *Comprehensive Treatise on Inorganic and Theoretical Chemistry* (London: Longmans, 1927), 7:305.

13. SULPHATE OF BARIUM AND FLUORIDE OF CALCIUM

149. "earlier in Derbyshire, England." James Smithson, "An Account of a Native Combination of Sulphate of Barium and Fluoride of Calcium," *Annals of Philosophy* 16 (1820): 48.

150. "barium must have been in the stone." Ibid., 49.

150. "having produced the new mineral." Ibid., 48–49; G. F. Loughlin and A. H. Koschmann, *Geology and Ore Deposits of the Magdalena Mining District, New Mexico* (Washington, DC: US Government Printing Office, 1942), 110, 113; Maximilian Keim, Benjamin F. Walter, Udo Neumann, Stefan Kreissl, Richard Bayerl, and Gregor Markl, "Polyphase Enrichment and Redistribution Processes in Silver-Rich Mineral Associations of the Hydrothermal Fluorite-Barite-(Ag-Cu) Clara Deposit, SW Germany," *Mineralium Deposita* 54 (2019): 155.

151. "he described that process in his article." James Smithson, *The Scientific Writings of James Smithson*, ed. William J. Rhees (Washington, DC, 1879), 73. Smithson's belief in binary combination was discussed in chapters 4 and 6.

151. "nearly seventeen years earlier." Smithson, "Barium and Fluoride," 50. This was discussed in chapters 4 and 6.

152. "chemical combination is binary." Smithson, *The Scientific Writings of James Smithson*, ed. William J. Rhees (Washington, DC, 1879), 73.

153. "included in that discussion." *Journal de Physique* 92 (1821): 77. For example, *Annaes das Sciencias* 14 (1821): 62; Th. Scheerer and E. Drechsel, "Schwespath und Flusspath, künstliche Darstellung derselben," *Journal für Praktische Chemie* 7 (1873): 63. An abstract of this article can be found in *Journal of the Chemical Society* 27 (1874): 234; Friedrich Naumann, *Elemente der Mineralogie* (Leipzig, 1874), 261. (My thanks to Jeff Gorman for this translation.) See also Reinhard Blum, "Flusspath," *Lehrbuch der Mineralogie* (Stuttgart, 1874), 170. For example, U.S. Patent 1752244A, "Process for Refining and Purifying Barium Sulphate," March 25, 1930.

153. "a test that, in his experience, would not work." James Smithson, "A Means of Discrimination between the Sulphates of Barium and Strontium," *Annals of Philosophy* 5 (1823): 359. "Miscellaneous Intelligence: Chemical Science," *Quarterly Journal* 10 (1821): 189. The citation of this notice in Smithson's article is incorrect.

153. "using 'wet' methods in their analysis." Smithson, "Barium and Calcium," 49.

154. "and so the test would not work." Smithson, "Sulphates of Barium," 359–360.

154. "strontia was not soluble in water." "Chemical Science," *Quarterly Journal* 15 (1823): 383; *Archives des découvertes: Pendant l'année 1823* (Paris, 1824), 137; *Bulletin des Sciences Mathématiques* 13 (1830): 369. For example, Ebenezer Emmons, *Manual of Mineralogy and Geology: Designed for the Use of Schools* (Albany, 1826), 85; Thomas Graham, *Elements of Inorganic Chemistry*, 2nd ed. (Philadelphia, 1858), 406.

155. "using the test successfully and with excellent results." Hope's work on strontia was finally published in the *Transactions of the Royal Society of Edinburgh* 4 (1798), 4. J. Andrews, "On the Detection of Baryta or Strontia when in union with Lime," *Philosophical Magazine* 7 (1830): 406.

155. "discussions of strontium's solubility." J. W. Marden, "The Solubilities of the Sulfates of Barium, Strontium, Calcium and Lean in Ammonium Acetate Solutions at 25° and a Criticism of the Present Methods for the Separation of these Substances by means of Ammonium Acetate Solutions," *Proceedings of the American Chemical Society for the Year 1916* (Easton, PA: ACS, 1916), 311.

14. A NEW TEST FOR ARSENIC AND "SMITHSON'S PILE"

157. "new methods to detect each of them." James Smithson, "On the Detection of Very Minute Quantities of Arsenic and Mercury," *Annals of Philosophy* 20 (1822): 127–128; James Smithson, "On a Native Compound of Sulphuret of Lead and Arsenic," *Annals of Philosophy* 14 (1819): 96–97.

157. "essentially unregulated and readily available." Mathieu Orfila, *A General System of Toxicology: or, a Treatise on Poisons* (Philadelphia: Carey, 1817), 46.

158. "death could result from as little as two." The modern equivalent to a grain is about 64.8 mg. James C. Whorton, *The Arsenic Century: How Victorian Britain Was Poisoned at Work, Home, and Play* (Oxford: Oxford University Press, 2010), 10.

158. "how it was administered." Ibid., 66–68.

159. "the doctor was charged with murder." Ibid., 83–85.

159. "characteristic color when arsenic was present." Ian A. Burney, "Bones of Contention: Mateu Orfila, Normal Arsenic and British Toxicology," in *Chemistry, Medicine, and Crime*, ed. José Ramón Bertomeu-Sánchez and Agustí Nieto-Galan (Sagamore Beach, MA: Science History Publications, 2006), 247–248.

159. "have a similar weakness." Whorton, *Arsenic Century*, 84–85; Cooper, *Tracts on Medical Jurisprudence* (London, 1819), 432.

160. "the colors of the tests' precipitates." John White Webster, *A Manual of Chemistry, on the Basis of Professor Brande's* (London, 1828), 395–397. Burney, "Bones of Contention," 248.

160. "in its metal-like elemental form." "Premiums Offered in the Session 1821–1822: Test for Arsenic," *Transactions of the Society for the Encouragement of Arts, Manufactures, and Commerce* 39 (1821): xiii.

161. "give a false positive for the poison." R. Phillips, "On the Methods of Employing the Various Tests Proposed for Detecting the Presence of Arsenic," *Annals of Philosophy* 7 (1824): 33–35.

161. "poisoning was a significant concern." Ibid., 35; "Tests for Arsenic," *The Chemist* 1 (1824): 12–14. Robert Hare, *Compendium of the Course of Chemical Instruction in the Medical*

Department of The University of Pennsylvania (Philadelphia, 1828); Theodric Beck, *Elements of Medical Jurisprudence* (London, 1825), 406; E. Littell, *The Museum of Foreign Literature, Science and Art* 1 (1822): 186.

161. "it had been largely forgotten." James Marsh, "Separation of Arsenic," *Transactions of the Society for the Encouragement of Arts, Manufactures, and Commerce* 51 (1836): 66–76; Henry Trueman Wood, *A History of the Royal Society of Arts* (London, 1913), 282; Whorton, *The Arsenic Century*, 86.

162. "while limiting its undesirable effects." Andrew Mathias, *An Inquiry into the History and Nature of the Disease Produced in the Human Constitution by the Use of Mercury* (Philadelphia, 1811), iv.

162. "residents with low incomes." André Guillerme, "Le mercure dans Paris: Usages et nuisances (1780–1830)," *Urban History* 1 (2007): 80.

163. "utilized the newly discovered 'galvanic force.'" Smithson, *The Scientific Writings of James Smithson*, ed. William J. Rhees (Washington, DC, 1879), 77.

164. "new shops and businesses opening every year." Ibid., 77–95.

165. "more than eight feet away." This description of hat making draws on Michael Sonenscher, *The Hatters of Eighteenth Century France* (Berkeley: University of California Press, 1987), 41.

165. "by just one trade." Gilders used even more mercury per capita. Guillerme, "Le mercure dans Paris," 87–88, 90, 93.

165. "was now dramatically increased." Ibid., 80, 85–86.

166. "heavy metals, especially mercury." Ibid., 93, 95; Pierre Chardon, "On the Acrodynia," *Medico-Chirurgical Review* 14 (1831): 200–202.

166. "hang over that part of the city." David S. Barnes, *The Great Stink of Paris and the Nineteenth-Century Struggle against Filth and Germs* (Baltimore: Johns Hopkins University Press, 2006), 76; Patrice Moncan, *Le Paris d'Haussman* (Paris: Du Mecene, 2009), 10. Period writers like Balzac mention the smoky cloud over the right bank.

166. "recently discovered galvanic force." *The Medical Times* 2 (1840): 59.

167. "unfortunately never published." René Just Haüy, *Extrait d'un traité élémentaire de minéralogie* (Paris, 1797), 5:89.

167. "their power would be combined." Henry Schlesinger, *The Battery* (Washington, DC: Smithsonian Books, 2010), 45; Alexander Volta, "On the Electricity Excited by the Mere Contact of Conducting Substances of Different Kinds," *Philosophical Magazine* 7 (1800): 311. Image from Adolphe Ganot, *Elementary Treatise on Physics*, 4th ed. (New York, 1893), 795.

167. "its ability to move matter." See Peter Mark Roget, "Galvanism," *Encyclopaedia Metropolitana* (London, 1830), 2:211.

168. "an entire chemical theory." Jöns Jakob Berzelius, *An Attempt to Establish a Pure Scientific System of Mineralogy, by the Application of the Electro-Chemical Theory and the Chemical Proportions* (London, 1814).

169. "the development of electroplating." Alexander Jamieson, *A Dictionary of Mechanical Science, Arts, Manufactures, and Miscellaneous Knowledge* (London, 1829), 1:351; "Galvanism," 211.

169. "covered with a layer of mercury." Smithson, "On the Detection," 128.

170. "engaging with some of the leading electrical researchers." James Smithson, "On Some Capillary Metallic Tin," *Annals of Philosophy* 17 (1821): 271; Oersted to his wife, March 5, 1832; Oersted Papers, Danish Royal Library, Copenhagen; cited in Heather Ewing, *The Lost World of James Smithson* (New York: Bloomsbury, 2007), 283–284.

170. "reported only the mercury test." *Polytechnisches Journal* 9 (1822): 266; *Polytechnisches Journal* 18 (1825): 486; William Henry, *The Elements of Experimental Chemistry*, 9th edition (Philadelphia and London, 1823), 2:571; *Annales de Chimie* 21 (1822): 97; *Archives des découvertes . . . Pendant l'année 1822* (Paris, 1823): 158; *Medical Jurisprudence* 2 (1823): 277; "Manière de découvrir de très-petites quantités de mercure," *Annales des Mines* 8 (1823): 175.

170. "was 'corrosive sublimate' ($HgCl_2$)." Charles Sylvester, "Experiments on the Decomposition of the Fixed Alkalis by Galvanism," *Journal of Natural Philosophy* 19 (1808): 156–157; Charles Sylvester, "On the Nature and Detection of the different Metallic Poisons," *Journal of Natural Philosophy* 33 (1812): 306–313.

170. "Sylvester's article." Charles Sylvester, "On the Nature and Detection of Different Metallic Poisons," *The Eclectic Repertory* 4 (1814): 449–456; Beck, *Elements of Medical Jurisprudence*, 445–446; Fredrick Accum, *A Practical Treatise on the Use and Application of Chemical Tests* (London, 1829), 303; Edmund Davy, "On a Simple Electro-Chemical Method of Ascertaining the Presence of Different Metals; Applied to Detect Minute Quantities of Metallic Poisons," *Philosophical Transactions* 121 (1831): 156.

171. "the situation elsewhere, particularly in France." See, e.g., Edward Turner, *Elements of Chemistry*, 5th ed. (Philadelphia, 1835), 398; Robert Christison, *A Treatise on Poisons*, 3rd ed. (Edinburgh, 1836), 347; Thomas Thomson, *A System of Chemistry* (London, 1831), 1:613.

171. "*North American Medical and Surgical Journal.*" Orfila's concern was that any tin in the test solution would be deposited on the positive terminal and resemble mercury. Mateu Orfila, "Sur le procédé proposé par M. James Smithson, pour découvrir de très-petites quantités de sublime corrosive ou d'un sel mercuriel," *Journal de Chimie Médicale* 6 (1829): 265. *Annales de Chimie et de Physique* 41 (1829): 92; *Annales des Mines* 1 (1832): 124–125; *The Philosophical Magazine* 6 (1829): 394–395; *The Quarterly Journal* 28 (1830): 183; *North American Medical and Surgical Journal* 8 (1829): 432–433.

171. "references to Sylvester ceased to appear." See, e.g., *Dictionnaire de médecine* (Paris, 1839), 19:587–588; *Journal de Chimie Médicale* 5 (1839): 200; C. Favrot, *Traité élémentaire de physique, chimie, toxicologie et pharmacie* (Paris, 1841), 1:529; "Recherches toxicologiques sur l'empoisonnement par le mercure," *Journal de Chimie Médicale* 11 (1845): 241–242; "Toxicological Investigations Concerning Poisoning by Mercury," *The Chemist* 6 (1845): 280; *The Medical Times* 12 (1845): 60–62.

172. "the use of 'Smithson's process.'" See, e.g., "Séances de l'Académie royale des sciences," *Archives Générales de Médecine* 8 (1845): 114–115; Otto Erdmann and Gustav Werther, "Ueber die Wirkung der Smithson'schen Kette bei der Untersuchung auf kleine Mengen Quechsilber," *Journal für Praktische Chemie* 86 (1862), 245–247; Alfred Joseph Naquet, *Legal Chemistry: A Guide to the Detection of Poisons* (New York, 1876), 36–39; George B. Wood, *The Dispensatory of the United States* (Philadelphia, 1865), 455.

172. "only English-language journals were consulted." James Smithson, *The Scientific Writings of James Smithson*, ed. William J. Rhees (Washington, DC, 1879), iii, 133, 144.

172. "as 'le docteur Smithson.'" "La méthode indiquée dans le *Journal de Chimie Médicale*, par le docteur Smithson," *Journal de Chimie Médicale* 5 (1839): 200.

15. SMITHSON'S LAMP AND THE "SAPPARE"

175. "such a commonplace topic." James Smithson, "Some Improvements of Lamps," *Annals of Philosophy* 20 (1822): 363–364.

176. "furnished with such long wicks." For example, Samuel Frederick Gray, *The Operative Chemist; Being a Practical Display of the3 Arts and Manufactures Which Depend upon Chemical Principles* (London, 1828), 191.

177. "could easily be replaced." Smithson, *The Scientific Writings of James Smithson*, ed. William J. Rhees (Washington, DC, 1879), 78.

178. "reaction to the article was tepid at best." For example: *Traité pratique de chimie appliquée aux arts et manufactures* (Paris, 1828), 1:340; Samuel Frederick Gray and Arthur Livermore Porter, *The Chemistry of the Arts: Being a Practical Display of the Arts and Manufactures Which Depend on Chemical Principles* (Philadelphia, 1830), 157–159; Charles F. Partington, *The British Cyclopaedia of the Arts and Sciences* (London, 1835), 2:906; *The Family Magazine; or Monthly Abstract of General Knowledge* 3 (1838): 15.

179. "'A Method of Fixing Particles on the Sappare.'" Saussure, "Nouvelles recherches," 3–4: Smithson, "Sappare," 412.

180. "used to identify and analyze minerals." James Smithson, "A Method of Fixing Particles on the Sappare," *Annals of Philosophy* 5 (1823): 412.

180. "'at the expense of its existence.'" Ibid.

180. "holder that Smithson may also have used." Horace de Saussure, "Nouvelles recherches sur l'usage du chalumeau dans la minéralogie," *Journal de Physique* 2 (1794): 3–44; James Macie, "An Account of Some Chemical Experiments on Tabasheer," *Philosophical Transactions* 81 (1791): 373.

181. "even better materials and methods." J. C. Delamétherie, "Extrait d'une lettre de M. James Smithson à J. C. Delamétherie," *Journal de Physique* 60 (1805): 179.

181. "quickly published in its journal." "Supports for Ignition of Particles by the Blow-pipe," *Quarterly Journal of Science, Literature and the Arts* 16 (1823): 370; "Sur un mode perfectionné de maintenir des petites particules de minéraux," *Bulletin des Sciences Technologiques* 1 (1824): 152; "M. Smithson. Sur quelques perfectionnemens dans les essays au chalumeau," *Annales des Mines* 12 (1826): 133; J. G. Totten, "Notes on Some New Supports for Minerals," *Annals of the Lyceum of Natural History of New York* 1 (1824): 109.

182. "in his textbook *Chemical Manipulation* (1842)." J. J. Berzelius, *The Use of the Blowpipe in Chemistry and Mineralogy*, trans. J. D. Whitney (Boston, 1845), 7, 21–22; Michael Faraday, *Chemical Manipulation: Being Instructions to Students in Chemistry*, 3rd ed. (London, 1842), 118–119.

182. "what would come to be known as microchemistry." Smithson, "Sappare," 413.

16. AN "ARISTOCRATIC SCIENCE DABBLER"?

183. "context in which he wrote them?" Nina Burleigh, *The Stranger and the Statesman* (New York: Perennial, 2004), 161–163.

183. "the topic's scientific literature." James Smithson, "On the Crystalline Form of Ice," *Annals of Philosophy* 21 (1823): 340. The article Smithson read was Jean André de Luc, "Des glacières naturelles, et de la cause qui forme la glace dans ces cavités," *Annales de Chimie et de Physique* 21

(1822): 125. The author was the nephew of Smithson's critic of the same name, who had died almost six years earlier. Smithson also made reference to Hericart de Thury, "Sur la cristallisation de la glace," *Journal des Mines* 33 (1813): 157, and Edward Daniel Clarke, "Upon the Regular Crystallization of Water," *Transactions of the Cambridge Philosophical Society* 1 (1822): 209–215. He also appears to have read César Despretz, "Sur la forme cristalline de la glace," *Annales de Chimie et de Physique* 21 (1822): 155–158. With regard to octahedral crystals, he refers to René Just Haüy, *Traité élémentaire de physique* (Paris, 1803), 1:172–173; René Just Haüy, *Traité élémentaire de physique*, 3rd ed. (Paris, 1821), 1:257; and M. D'Antic, "Observations sur la cristallisation de la glace," *Journal de Physique* 33 (1788): 56–59.

184. "'. . . perfectly formed in hexagons.'" Johannes Kepler, *On the Six-Cornered Snowflake* (1611). The quotation from Descartes is taken from F. C. Frank, "Descartes' Observations on the Amsterdam Snowfalls of 4, 5, 6 and 9 February 1634," *Journal of Glaciology* 13 (1974): 535.

184. "'opinions accurate?' he asked." Smithson, "On the Crystalline Form of Ice," 340.

184. "crystals in these caves existed year-round." Clarke, "Upon the Regular Crystallization of Water," 212–213.

184. "'two hexagonal pyramids joined base to base.'" Smithson, "On the Crystalline Form of Ice," 340.

185. "examples of crystals with this property." Unpublished study referred to in Haüy, *Extrait d'un traité élémentaire de minéralogie* (Paris, 1797), 89; Smithson, "On the Crystalline Form of Ice," 340.

185. "having different temperatures." J. Latham and B. J. Mason, "Generation of Electric Charge Associated with the Formation of Soft Hail in Thunderclouds," *Proceedings of the Royal Society of London, Series A, Mathematical and Physical Sciences* 260 (1961): 537; D. Müller-Hillebrand, "Charge Generation in Thunderstorms by Collision of Ice Crystals with Graupel, Falling through a Vertical Electric Field," *Tellus* 6 (1954): 367.

185. "he was widely credited for it." "Sur la forme cristalline de la glace, par James Smithson," *Bulletin Général et Universel des Annonces et des Nouvelles Scientifiques* 2 (1823): 420; *Arsberättelser om Vetenskapernas Framsteg* 4 (1824): 84; Jacob Berzelius, *Jahres-Bericht über die Fortschritte der physischen Wissenschaften* (Tubingen, 1824), 75; "Krystallform des Eises," *Jahrbucher des Kaiserlichen Koniglichen Polytechnischen Institutes in Wien* (Vienna, 1825), 6:421–422; M. Bauer, W. Dames, and T. Liebisch, *Neues Jahrbuch für Mineralogie, Geologie und Palaeontologie* (Stuttgart, 1886), 2:184–186; M. G. Delafosse, *Nouveau cours de minéralogie* (Paris, 1862), 3:89; James Fowler, *Water: Its Nature and Natural Varieties* (London, 1865), 15. See also Thomas Thomson, *System of Chemistry of Inorganic Bodies*, 7th ed. (London, 1831), 1:103.

186. "indulging in an obscure topic." Smithson, "On the Crystalline Form of Ice," 340.

186. "between coffee and English science." James Smithson, "An Improved Method of Making Coffee," *Annals of Philosophy* 6 (1823): 30–31; Leonard Carmichael and J. C. Long, *James Smithson and the Smithsonian Story* (New York: Putnam, 1965), 113–115; Burleigh, *The Stranger & the Statesman*, 159.

186. "throughout the eighteenth century." Aytoun Ellis, *The Penny Universities: A History of the Coffee-Houses* (London: Secker and Warburg, 1956), 18–24; see also "John Wilkins" in John Aubrey, *Aubrey's Brief Lives*, ed. Oliver Lawson Dick (Ann Arbor: University of Michigan Press, 1962).

186. "was also a Grecian regular." William Harrison Ukers, *All About Coffee* (New York: The Tea and Coffee Trade Journal Co., 1922), 74; John Timbs, *Club Life of London* (London: Bentley, 1866), 2:202–203.

186. "all took place in London coffeehouses." Christa Jungnickel and Russell McCormmach, *Cavendish* (Philadelphia: American Philosophical Society, 1996), 70; Albert Edward Musson and Eric Robinson, *Science and Technology in the Industrial Revolution* (New York: Routledge, 1994), 126–127.

187. "to join the scientific community." Heather Ewing, *The Lost World of James Smithson* (New York: Bloomsbury, 2007), 77–78, 109.

187. "he was elected to the Royal Society." Trevor Levere and Gerard L'E Turner, *Discussing Chemistry and Steam* (Oxford: Oxford University Press, 2002).

187. "coffee actually tasted very good." Count Rumford, *The Complete Works of Count Rumford* (London: Macmillan, 1876), 5:618; John Ellis, *An Historical Account of Coffee* (London, 1774), 28, 39; Ellis was a member of the Royal Society.

187. "that much hotter." Ukers, *All About Coffee*, 696–697. Isinglass was made exclusively from Russian sturgeon until 1795, when a process was developed to make it from cod. Rumford, *Complete Works*, 5:633.

189. "his standard unit of coffee." Rumford, *Complete Works*, 5:620–622, 627–628, 631–632, 657. Rumford's measurements were all in English inches, French inches being slightly different at this time.

190. "other possible uses." James Smithson, "An Improved Method of Making Coffee," 30–31.

17. CHLORIDE OF POTASSIUM

191. "he called the article." James Smithson, "A Discovery of Chloride of Potassium in the Earth," *The Annals of Philosophy* 6 (1823): 258–259.

192. "the unknown salt." The use of a matrass is described in Abraham Rees, "Phosphorus of Sulphur," in *Cyclopaedia; or Universal Dictionary of Arts, Sciences, and Literature* (London, 1819), vol. 27; Thomas Thomson, *System of Chemistry* (London, 1818), 1:188.

192. "did not contain sulfur." Chloride of barium was also used as an investigative tool by other chemists, notably Thomas Thomson. See Edward Turner, "On the Composition of Chloride of Barium," *Philosophical Transactions* 119 (1829): 292.

193. "as chloride of potassium (KCl)." This test is described in Richard Phillips, "On the Action of Chlorides and Water," *Annals of Philosophy* 1 (1821): 31, and Franklin Bache, *A System of Chemistry for the Use of Students of Medicine* (Philadelphia, 1819), 307–308. Note that "crystals of tartar" was an alternative name for "cream of tartar."

193. "to confirm Smithson's analysis." This was a well-known test. See George Fownes, *Elementary Chemistry, Theoretical and Practical* (Philadelphia, 1847), 275; Arnold James Cooley, "Platinum," in *Cooley's Cyclopaedia of Practical Receipts and Collateral Information*, 6th ed. (London, 1880), 2:1339.

193. "separated from that insight." "Miscellaneous Intelligence," *The Quarterly Journal of Science, Literature, and the Arts* 16 (1823): 395; "Kalium-Chlorid. James Smithson," *Jahrbücher des Kaiserlichen Königlichen Polytechnischen Institutes in Wien* 6 (1825): 322; *Zeitschrift für*

Mineralogie 2 (1825): 518; "Mineralogie," *Magazin für Pharmacie und die dahin einschlagenden* 9–10 (1825): 229; F. S. Beudant, *Traité élémentaire de minéralogie* (Paris, 1832), 511. Ironically, Beudant was the first to propose that zinc carbonate be named "Smithsonite." For the citation of Smithson, see, for example, James D. Dana, *A System of Mineralogy, Comprising the Most Recent Discoveries*, 4th ed. (New York, 1854), 1:90.

194. "he chose to remain silent." Humphry Davy, "Researches on the Oxymuriatic Acid. Its Nature and Combinations: And on the Elements of the Muriatic Acid. With Some Experiments on Sulphur and Phosphorus," *Philosophical Transactions* 100 (1810): 231–257. For more on the chloridic theory, see John Ayrton Paris, *The Life of Sir Humphry Davy* (London, 1831), 210–214; Richard Phillips, "On the Action of Chlorides and Water," *Annals of Philosophy* 1 (1821): 27–33. By the 1830s this theory had been widely accepted.

18. COMPOUNDS OF FLUORINE

195. "*fluere*, meaning 'to flow.'" James Smithson, "On Some Compounds of Fluorine," *Annals of Philosophy* 7 (1824): 100–101.

196. "the acid ate holes in them." "101. Black to James Watt: Edinburgh, 23 December 1772," *The Correspondence of Joseph Black*, ed. Robert Anderson and Jean Jones (Edinburgh: Ashgate, 2012), 1:264–265, 377–379; Joseph Priestley, "Of the Fluor Acid Air," *Experiments and Observations on Different Kinds of Air* (London, 1776), 2:197.

196. "'taking the form of air.'" Thomas Beddoes, ed., *The Chemical Essays of Charles-William Scheele* (London: Scott, Greenwood & Co., 1901), xxi–xxii; "176. Black to James Watt: Edinburgh, 7 January 1780," *The Correspondence of Joseph Black*, 1:408–409; Priestley, *Experiments and Observations*, 188.

196. "he named it 'fluoric.'" Martin Wall, *A Syllabus of a Course of Lectures in Chemistry* (Oxford, 1783), 12, 15; Antoine Lavoisier, *Elements of Chemistry*, trans. Robert Kerr (Edinburgh, 1790), 175–176, 240–241.

196. "experiments of his own." "James Smithson mineral catalogue," Smithsonian Institution Archives, RU 7000, Box 2; Priestley, *Experiments and Observations*, 187, 189; Lavoisier, *Elements of Chemistry*, 240; Humphry Davy, "An Account of Some New Experiments on the Fluoric Compounds: With Some Observations on Other Objects of Chemical Inquiry," *Philosophical Transactions* 104 (1814): 64; Thomas Thomson, *A System of Chemistry* (London, 1810), 215.

197. "a second article on the same topic." Humphry Davy is frequently credited with introducing this name, but he credited Ampère with suggesting it. Humphry Davy, "Some Experiments and Observations on the Substances Produced in Different chemical Processes on Fluor Spar," *Philosophical Transactions* 103 (1813): 278; N. C. Datta, *The Story of Chemistry* (Hyderabad: Universities Press, 2005), 428; John Davy, "An Account of Some Experiments on Different Combinations of Fluoric Acid," *Philosophical Transactions* 102 (1812): 352. Davy, "Some Experiments," 267; Davy, "An Account of Some New Experiments on the Fluoric Compounds," 62.

197. "needing further investigation." James Smithson, "On the Composition of Zeolite," in *The Scientific Writings of James Smithson*, ed. William J. Rhees (Washington, DC, 1879), 46; Davy, "Some Experiments and Observations," 279; Smithson, "Compounds," 103. It was the French chemist Ampère who first suggested that hydrogen combined with fluoride, which Davy acknowledged in their correspondence. See their letters from 1812–14.

198. "danger of working with this strange substance." James Smithson, "An Account of a Native Combination of Sulphate of Barium and Fluoride of Calcium," *Annals of Philosophy* 16 (1820): 48. "Smithson's Library," Smithsonian Libraries, contains pamphlets inscribed by Gay-Lussac to "Monsieur de Smithson."

198. "he presented one." Smithson, "Compounds," 101, 103; Charles Eliot, *A Compendious Manual of Qualitative Chemical Analysis* (New York, 1869), 93.

198. "'their nature may be ascertained.'" J. J. Berzelius, *The Use of the Blowpipe in Chemical Analysis* (London, 1822), 35, plate III, fig. 2.

198. "served the same purpose." J. J. Berzelius, *The Use of the Blowpipe*, trans. J. D. Whitney (Boston, 1845), 83–84; Smithson, "Compounds," 102. The use of "indicator" papers was discussed in Chapter 10.

199. "the hollow tube, where they could be studied." J. J. Berzelius, *The Use of the Blowpipe in Chemical Analysis* (London, 1822), 33.

199. "to protect it." Smithson, "Compounds," 101–102; James Smithson, "A Method of Fixing Particles on the Sappare," *Annals of Philosophy* 6 (1823): 412; Berzelius, *The Use of the Blowpipe*, 84.

200. "gases from the sample into the glass tube." Smithson, *The Scientific Writings of James Smithson*, ed. William J. Rhees (Washington, DC, 1879), 78.

200. "other, more quantitative, tests." Abraham Rees, "Fluor," *The Cyclopaedia, or, Universal Dictionary of Arts, Sciences, and Literature,* 14 (London: Longman, 1819); "Sur quelques composés de fluor; par M. Smithson," *Bulletin des Sciences Mathématiques* 1 (1824): 162; Leopold Gmelin, *Hand-Book of Chemistry*, trans. Henry Watts (London, 1849), 3:213; J. J. Berzelius, "Fluorures," *Traité de chimie* (Paris, 1831), 3:341; Berzelius, *The Use of the Blowpipe*, 83–84; J. J. Berzelius, *Die Anwendung des Othrohrs* (Nuremburg, 1844); 106. Thomas Richter, "Fluorine," *Plattner's Manual of Qualitative and Quantitative Analysis with the Blowpipe*, 5th ed. (New York, 1885), 4, 377; Ulrich Burchard, "The History and Apparatus of Blowpipe Analysis," *The Mineralogical Record* 25 (1994): 259; Smithson, *Scientific Writings*, 94.

201. "tests like a gift." Smithson, "Compounds," 100–101, 103–104.

19. EGYPTIAN COLORS

203. "'and in many of the arts.'" James Smithson, "An Examination of Some Egyptian Colours," *Annals of Philosophy* 23 (1824): 115–116; Rob Iliffe, *Priest of Nature: The Religious Worlds of Isaac Newton* (Oxford: Oxford University Press, 2017), 189–192, 196–201, 204–205, 208–211; Martin Wall, *Dissertations on Select Subjects in Chemistry and Medicine* (Oxford, 1783), 27.

204. "would return to the topic many years later." In his article Smithson mentions the figure of Isis as coming from a "relation." Ewing identifies this as Henry Louis Dickenson, Smithson's brother, who had apparently been among the early English tourists. Dickenson returned to London early in 1803, after a long trip through Europe. Smithson, "An Examination of Some Egyptian Colours," 116; Heather Ewing, *The Lost World of James Smithson* (New York: Bloomsbury, 2007), 215.

204. "Giovanni Battista Belzoni." Tom Verde, "Egyptology's Pioneering Giant," *AramcoWorld* (2018), 30.

204. "massive shoulders and a barrel chest." Stanley Mayes, *The Great Belzoni: The Circus Strongman Who Discovered Egypt's Ancient Treasures* (London: Tauris, 2003), 17–42.

204. "found the lost city of Berenice." Ibid., 11, 116.

205. "an associate, James Curtin." Giovanni Belzoni, *Catalogue of the Various Articles of Antiquity, to Be Disposed of at the Egyptian Tomb, by Auction, or by Private Contract* (London, 1822). Image from plate IV. Dr. Robert Morkot, "The 'Irish lad' James Curtin, 'servant' to the Belzonis," ASTENE *Bulletin* 56 (2013): 16–19, with an additional note in *Bulletin* 65 (2015): 18; Mayes, *The Great Belzoni*, 281.

205. "inspired his article." Smithson, "Examination," 115–116.

206. "as the wax was being removed." Giovanni Belzoni, *Researches and Operations of G. Belzoni in Egypt and Nubia* (London: 1820).

207. "The red pigment was iron oxide." James Smithson, "A Chemical Analysis of Some Calamines," *Philosophical Transactions* 93 (1803): 13. He also mentioned using this test in "Account of a Discovery of Native Minium." Davy also used it: "Some Experiments and Observations on the Colours Used in Painting by the Ancients," *Philosophical Transactions* 105 (1815): 101.

208. "as the particle size increased." Terence E. Warner, *Synthesis, Properties and Mineralogy of Important Inorganic Materials* (New York: Wiley & Sons, 2011), 39–41. "Egyptian blue" is now known as calcium copper silicate or cuprorivaite ($CaCuSi_4O_{10}$). G. D. Hatton, A. J. Shortland, and M. S. Tite, "The Production Technology of Egyptian Blue and Green Frits from Second Millennium BC Egypt and Mesopotamia," *Journal of Archaeological Science* 35 (2008): 1591; T. Katsaros, Ioannis Liritzis, and N. Laskaris, "Identification of Theophrastus' Pigments *eyptios yanos* and *psimythion* from Archaeological Excavations," *Archeo Sciences* 34 (2010): 69–80.

208. "around 500 CE." Warner, *Synthesis, Properties and Mineralogy of Important Inorganic Materials*, 35.

209. "all of whom Smithson knew." Martin Klaproth, "Sur quelques vitrifications antiques," *Mémoires de l'Académie Royale des Sciences et Belles-Lettres* (Berlin, 1801), 3–16; Jean-Antoine Chaptal, "Sur quelques couleurs trouvées à Pompeia," *Annales de Chimie* 70 (1809): 22–31; Humphry Davy, "Some Experiments and Observations on the Colours Used in Painting by the Ancients," 97–124.

209. "of more than just passing interest." *Description de l'Égypte: ou, Recueil des observations et des recherches*, 2nd ed. (Paris, 1821), 3:100–101. Two chemists analyzed the blue pigment found on the sarcophagus of Seti I, and both found that it contained copper. However, the condition of the pigment, the small size of the sample, and the likelihood that it had been contaminated made these analyses less than definitive. Edward Daniel Clarke, "On the Chemical Examination, Characters, and Natural History of Arragonite," *Annals of Philosophy* 2 (1821): 57; J. G. Children, "On the Nature of the Pigment in the Hieroglyphics on the Sarcophagus from the Tomb of Psammus," *Annals of Philosophy* 2 (1821): 389.

209. "had also written on this topic." "Examen de quelques couleurs égyptiennes, par James Smithson," *Bulletin des Sciences Technologiques* 1 (1824): 222–223; "Untersuchung einiger Farben der alten Aegypter, Von Hrn. Jak Smithson," *Polytechnisches Journal* 13 (1824): 518; H. de Fontenay, "Sur le bleu égyptien," *Comptes Rendus* 78 (1874): 908. Fontenay's work was discussed in the Smithsonian Institution's *Annual Record of Science and Industry for 1874* (New York, 1875), 536.

20. KIRKDALE CAVE AND PENN'S THEORY

211. "the Royal Institution's *Quarterly Journal.*" James Smithson, "Some Observations on Mr. Penn's Theory Concerning the Formation of the Kirkdale Cave," *Annals of Philosophy* 7 (1824): 50–60; "(Review of) Supplement to the Comparative Estimate of the Mineral and Mosaical Geologies, Relating Chiefly to the Geological Indications of the Phenomena of the Cave of Kirkdale," *Quarterly Journal of Science, Literature, and the Arts* 16 (1823): 309–321.

211. "the time in which he lived." Smithson, "Some Observations," 51. Smithson's 1813 article was discussed in chapter 9.

212. "allow him to make an analysis." Martin J. S. Rudwick, *Bursting the Limits of Time* (Chicago: University of Chicago Press, 2005), 622–632.

212. "this was what he meant." Ibid., 606–607. Buckland's beliefs are obvious from the titles of his previous and subsequent works: William Buckland, *Vindiciae Geologicae; or the Connexion of Geology with Religion Explained* (Oxford, 1820); William Buckland, *Reliquiae Diluvianae; of Observations on the Organic Remains Contained in Caves, Fissures, and Diluvial Gravel; and on Other Geological Phenomena Attesting the Action of an Universal Deluge* (London, 1823).

213. William Buckland, "Account of an assemblage of Fossil Teeth and Bones of Elephant, Rhinoceros, . . ." *Philosophical Transactions* 112 (1822): Plate 16.

214. "reported that he too had read it." Buckland, "Account of an Assemblage," 171–236; Rudwick, *Bursting*, 628, 631; Smithson, "Some Observations," 51.

214. "the Bible's description of a single great deluge." Gn 6:14–8:14; Ralph O'Connor, "Young-Earth Creationists in Early Nineteenth-Century Britain? Towards a Reassessment of 'Scriptural Geology,'" *History of Science* 45 (2007): 357–403; Nicolaas Rupke, *The Great Chain of History* (Oxford: Clarendon Press, 1983), 42–50; Granville Penn, *A Comparative Estimate of the Mineral and Mosaical Geologies* (London, 1822); Granville Penn, *Supplement to the Comparative Estimate of the Mineral and Mosaical Geologies* (London, 1823), 2.

215. "no matter how strong the evidence." Albert Edmunds, "Granville Penn as a Scholar," *Pennsylvania Magazine of History and Biography* 19 (1895): 119; Granville Penn, *A Comparative Estimate of the Mineral and Mosaical Geologies* (London, 1825) 2:285–286; Penn, *Comparative Estimate*, v–vii.

215. "called on him to address." "(Review of) Penn's Mineral and Mosaical Geologies," *Eclectic Review* 19 (1823): 37–53; "Analysis of Scientific Books," *The Quarterly Journal* 15 (1823): 108–109.

215. "felt compelled to oppose." "(Review of) Supplement to the Comparative Estimate," *Quarterly Journal* 16 (1823): 309.

215. "the origin of the bones." Ibid., 319.

216. "if one knew how to look." Rhoda Rappaport, "The Case of Noah's Flood in Eighteenth-Century Thought," *British Journal for the History of Science* 11 (1978): 2; Davis A. Young, *The Biblical Flood: A Case Study of the Church's Response to Extrabiblical Evidence* (Kalamazoo, MI: Eerdmans, 1995), 98–109; Nicolaas A. Rupke, *The Great Chain of History* (Oxford: Oxford University Press, 1983), 42–50.

216. "estrangement from the Royal Society." De Luc's influence was discussed in Chapter 9. J. L. Heilbron, "Historian of Nature and Mankind," in *Jean-André Deluc: Historian of Earth and Man* (Geneva: Slatkine, 2011), 100–103. See for example, J. A. Deluc, "Letters to Dr James Hutton on

the Theory of the Earth," *Monthly Review* 2 (1790): 206–227, 582–601; 3 (1790): 573–586; 5 (1791): 564–585; J. A. de Luc, "Remarks on the Geological Theory Supported by James Smithson," *Philosophical Magazine* 43 (1814): 127–137.

216. "aggressive tone of his article." "Literary and Scientific Notices," *The London Journal of Arts and Sciences* 6 (1823): 112.

217. "the mud hardened into limestone." "(Review of) Supplement," 310–312.

217. "'lived in any agitated sea.'" Smithson, "Some Observations," 52–56.

218. "'the suffocating effluvium.'" Ibid., 56; Laurence Irving, "The Precipitation of Calcium and Magnesium from Sea Water," *Journal of the Marine Biological Association* 14 (1926): 441. For a discussion of what was known about limestone formation in Smithson's time, see Sally Newcomb, "Contributions of British Experimentalists to the Discipline of Geology: 1780–1820," *Proceedings of the American Philosophical Society* 134 (1990): 161–225.

219. "Nothing like this had been found." Smithson, "Some Observations," 53–54.

219. "had been found in the Kirkdale cave." Ibid., 54.

219. "'all of them must have perished.'" M. Michael Adams, "Some Account of a Journey to the Frozen Sea, and of the Discovery of the Remains of a Mammoth," *Philosophical Magazine* 24 (1808): 141; Smithson, "Some Observations," 53–58.

220. "not a single one had ever been found." Gn 7:23; Smithson, "Some Observations," 58–60.

221. "'fish, or shell, or insect, or plant, is now alive.'" Gn 6:19.

221. "he never mentioned Smithson." In the second edition of his *Comparative Estimate* (1825), 1:xi, Penn acknowledged criticisms that appeared in the *Eclectic Review*, *Journal of Science* and the *British Critic*. Most of his responses to that criticism were concentrated in the introduction.

221. "'had a superficial knowledge of the subject.'" *The British Critic* 21 (1824): 387, 402; "Review of New Books," *The Literary Gazette and Journal of Belles Lettres, Arts, Sciences, etc.* 371 (1824): 133. For example: Anonymous, *Conversations on Geology* (London, 1828); Milton Millhauser, "The Scriptural Geologists," *Osiris* 11 (1954): 71. For an informed discussion of Penn's critics other than Smithson, see Walter Cannon, "The Problem of Miracles in the 1830s," *Victorian Studies* 4 (1960): 15–18.

222. "why he found it so objectionable." "Notice and Review of the 'Reliquiae Diluvianae,'" *American Journal of Science* 8 (1824): 154–157; "Sur la théorie de M. Penn; par M. Smithson," *Bulletin des Sciences Naturelles et de Geologie* 5 (1825): 25.

222. "first decades of the nineteenth century." Smithson, "Some Observations," 51 (emphasis added).

222. "suggestive, if not definitive." The intellectual climate in England in the early nineteenth century was discussed in chapter 9. "Notices," *The London Journal*, 112. The editor of the *Quarterly Journal* at this time was William Brande.

223. "context of the Kirkdale cave." Smithson, "Some Observations," 54, 57; F. B. Meek, "Remarks on the Carboniferous and Cretaceous Rocks of Eastern Kansas and Nebraska," *American Journal of Science and Arts* 39 (1865): 173–174; Rudwick, *Bursting*, 385–388, 604–609; Georges Cuvier, *Recherches sur les ossements fossiles de quadrupèdes*, new ed. (Paris, 1823), 4:299, 369. William Buckland, *Reliquiae Diluvianae*, 2nd ed. (London, 1824), 107.

223. "this topic had not changed." Heather Ewing, *The Lost World of James Smithson* (New York: Bloomsbury, 2007), 50–51. Smithson's letters from Paris and their religious implications were discussed in chapter 4. Two other scholars have studied this article and come to similar conclusions: Anonymous, "The Smithsonian Institute," *Southern Literary Messenger* 5 (1839): 830; W. H. Brock, "British Science Periodicals and Culture: 1820–1850," *Victorian Periodicals Review* 21 (1988): 47–55. The well-informed author of the first noted Smithson's "physico-theological opinions," a clear reference to natural theology. Brock, while misinterpreting Smithson's suggestion of a new miracle, nonetheless located him among the advocates of "natural law."

224. "revealed about Smithson himself." The sole exception was the summary in *Bulletin des Sciences Naturelles* 5 (1825): 25. Charles Pleydell Neale Wilton, *Remarks on Certain Parts of Mr Granville Penn's Comparative Estimate of the Mineral and Mosaical Geologies* (London, 1826); Penn, *A Comparative Estimate of the Mineral and Mosaical Geologies*, vol. 2 (1825); O'Connor, "Young-Earth Creationists," 373.

21. THE "INCREASE AND DIFFUSION OF KNOWLEDGE"

225. "'I here impart.'" James Smithson, "On the Discovery of Acids in Mineral Substances," *Annals of Philosophy* 5 (1823): 384; emphasis added.

226. "'the greatest modern chemists.'" Robert Hooper, *Lexicon-Medicum or Medical Dictionary*, 2nd ed. (New York, 1824), 11; Thomas Thomson, *Outlines of Mineralogy, Geology, and Mineral Analysis* (London, 1836), 1:4, 7–9. A period definition of "mineral" was a solid material from the Earth's crust that was either composed entirely of one substance or else a homogeneous chemical combination of substances that were always in the same proportions. Any other stony substance was considered a "mechanical mixture" and was a topic for geology. Smithson, "Discovery of Acids," 384.

226. "analyzing an unknown mineral." A contemporary discussion of the history of mineral analyses can be found in Thomson, *Outlines of Mineralogy*, 4–7.

226. "it did not contain an acid." Smithson, "Discovery of Acids," 384; Jacob Berzelius, *The Use of the Blowpipe in Chemistry and Mineralogy* (London, 1845), 33, 81.

227. "'Quantities of Arsenic and Mercury' (1822)." James Smithson, "An Account of a Native Combination of Sulphate of Barium and Fluoride of Calcium," *Annals of Philosophy* 16 (1820): 49; James Smithson, "On the Detection of Very Minute Quantities of Arsenic and Mercury," *Annals of Philosophy* 4 (1822): 127–128. Both tests were also mentioned in James Smithson, "On a Native Compound of Sulphuret of Lead and Arsenic," *Annals of Philosophy* 14 (1819): 97. The sulfuric acid test was discussed in chapter 14 and the arsenic acid test in chapter 16.

227. "just such a sample." Smithson, "Discovery of Acids," 385.

228. "'saline substances, held in solution.'" Davies Gilbert, "Addresses Delivered before the Royal Society," *Abstracts of the Papers Printed in the Philosophical Transactions of the Royal Society of London* 3 (1830): 9.

228. "in the *Annales des Mines* (1826)." Smithson, "Discovery of Acids," 387. For the sulfuric acid and hydrochloric acid tests, see, e.g., Friedrich Walchner, *Handbuch der gesammten Mineralogie* (Karlsruhe, 1829), 1:159, 307, 323, 344–345, 375, 409. For the arsenic test, see, e.g., *Polytechnisches Journal* 3 (1824): 497–498; *Bulletin Général et Universel des Annonces et des Nouvelles Scientifiques* 2 (1823): 418; "Nouvelles méthodes pyrognostiques pour reconnaitre la presence des acides minéraux dans des atomes," *Bulletin des Sciences Mathématiques* 2 (1824): 239; "Sur

quelques perfectionnements dans les essais au chalumeau; par M. Smithson," *Annales des Mines* 12 (1826): 133.

228. "'as the chemist could in several hours.'" Thomas Gill, *Technical Repository* (London, 1823), 3:382; H. Hodgson, "Thoughts on Mineralogical Systems," *Gleanings in Science* 1 (1829): 259.

228. "as a 'Classic of Science.'" Thomas Thomson, *Outlines of Mineralogy, Geology, and Mineral Analysis* (London, 1836), 2:400; J. J. Berzelius, *The Use of the Blowpipe in Chemistry and Mineralogy*, trans. J. D. Whitney (Boston, 1844), 81; J. J. Berzelius, *Lehrbuch der Chemie* (Dresden, 1841), 10:407; "Classics of Science," *The Science News-Letter* 15 (1929): 389.

229. "analyzing a tear." Sir Archibald Geikie, *Annals of the Royal Society Club* (London: Royal Society, 1917), 286.

229. "an old letter caught his eye." H. Hodgson, "Thoughts on Mineralogical Systems," *Gleanings in Science* 1 (1829): 259. See chapter 20.

230. "'not thicker than a shilling.'" James Smithson, "A Letter from Dr. Black Describing a Very Sensible Balance," *Annals of Philosophy* 10 (1825): 52.

230. "seems to have been true of Smithson." Smithson, *The Scientific Writings of James Smithson*, ed. William J. Rhees (Washington, DC, 1879), 118.

230. "all that was necessary." Robert G. W. Anderson and Jean Jones, eds., *The Correspondence of Joseph Black* (Edinburgh: Ashgate, 2012), 2:1095–99.

230. "made it ideal for traveling." Ibid., 2:52–53; Trevor Levere, "Sons of Genius: Chemical Manipulation and Its Shifting Norms from Joseph Black to Michael Faraday," *Bulletin for the History of Chemistry* 35 (2010): 4.

231. "lighter than any of the scale's weights." Ibid., 53.

231. "required using a 'scale pan.'" In his letter to Smithson, Black said "had I occasion for a more delicate one [balance], I could make it easily by taking a much thinner and lighter slip of wood, and grinding the needle to give it an edge. It would also be easy to make it carry small scales of paper for particular purposes." Ibid., 53. The evaluation of Black's balance appeared in Henry Kater and Dionysius Lardner, eds., *The Cabinet Cyclopaedia: Mechanics* (London, 1830), 293–294. When using the pans, the fulcrum needed to be placed on a raised support.

231. "served a mostly working-class audience." "Miscellaneous Intelligence," *The Quarterly Journal* 20 (1826): 161. It should be noted that the Royal Institution also had an interest in the education of working men. "On a Very Sensible Balance, for Weighing Small Globules of Metals," *The Technical Repository* 8 (1826): 76; *Glasgow Mechanics Magazine* 4 (1826): 154.

231. "'a good balance.'" Michael Faraday, *Chemical Manipulation, Being Instructions to Students in Chemistry* (London, 1827), 63–64; 3nd ed. (London, 1842), 62; Faraday, *Manipulations chimiques* (Paris, 1827), 1:73–77; Faraday, *Chemische Manipulation*, 75–79.

232. "included a description of Black's balance." Samuel Gray, *The Operative Chemist, Being a Practical Display of the Arts and Manufactures Which Depend on Chemical Principles* (London, 1828), title page, 233–234; Gray, *Traité pratique de chimie* (Paris, 1828), 1:409–413.

232. "no known examples survive." See, e.g., Anderson and Jones, *The Correspondence of Joseph Black*, 828–829. No original examples of Black's balance are known to survive, but a model, based on the letter in Smithson's article, was commissioned by the Royal Scottish Museum, Edinburgh, in 1971. It is now in the National Museum's collection, registration number 1971–272.

232. "both students and adult workers." Henry Violette, *Nouvelles manipulations chimiques simplifiées, ou laboratoire économique de l'étudiant ouvrage*, 3rd ed. (Paris, 1860), 20–21.

232. "of much interest to the scientific community." Speter has observed that if Black or Smithson had published a description of the balance prior to about 1818, it would have established a priority claim for the "horseman," or sliding weight—an important innovation for scales. Max Speter, "Joseph Black's 'Mikrowaage' mit Reiterversatz," *Zeitschrift für Instrumentenkunde*, 50 (1930): 204. But by the time the letter actually *was* published, this balance was of minimal interest to chemists.

233. "'which were bulky and heavy.'" Heather Ewing, *The Lost World of James Smithson* (New York: Bloomsbury, 2007), 297, 405n6; James Smithson, "A Method of Fixing Crayon Colours," *Annals of Philosophy* 10 (1825): 236.

233. "easy to mount in a protective frame." The colored-wax "crayons" so familiar in our time would not be introduced until the late 1800s.

233. "so the glass would not be needed." Marjorie Shelley, "The Rise of Pastel in the Eighteenth Century," https://www.metmuseum.org/blogs/now-at-the-met/features/2011/the-rise-of-pastel-in-the-eighteenth-century (last accessed October 29, 2019).

234. "the journal would have accepted it." James Smithson, "A Method of Fixing Crayon Colours," *Annals of Philosophy* 26 (1825): 236.

235. "'rendering the cold tints too predominant.'" Neil Jeffares, *Dictionary of Pastellists before 1800*, http://www.pastellists.com/Articles/Alexandre.pdf (last accessed October 29, 2019); *The Monthly Visitor, and Entertaining Pocket Companion* (London, 1797), 2:387; William Enfield, *Young Artist's Assistant: Or Elements of the Fine Arts*, 6th ed. (London, 1823), 222.

235. "'disposed to make trial of it.'" Smithson, "Crayon Colours," 236.

235. "he decided not to try." Ibid.; Tatyana Petukhova, "Potential Applications of Isinglass Adhesive for Paper Conservation," *The American Institute for Conservation, The Book and Paper Group Annual* 8 (1989): 58; Abigail B. Quandt, "Recent Developments in the Conservation of Parchment Manuscripts," *The American Institute for Conservation, The Book and Paper Group Annual* 15 (1996): nn. 15 and 16.

236. "forms have been identified." George Gregory, *A Dictionary of Arts and Sciences* (London, 1806), 346; Charles S. Tumosa and Marion F. Mecklenburg, "Oil Paints: The Chemistry of Drying Oils and the Potential for Solvent Disruption," in *New Insights into the Cleaning of Paintings* (Washington, DC: Smithsonian Books, 2013), 52.

236. "that noted Smithson's article." Charles Smith was a partner in a series of London art-supply firms, including Smith, Warner & Co. (1800–1820). See the online database of the National Portrait Gallery, "British artists' suppliers, 1650–1950," https://www.npg.org.uk/research/programmes/directory-of-suppliers/s (last accessed October 30, 2019).

236. "reprinted in a German drawing manual." "Bleistift-Zeichnungen auf Papier haltbar zu machen," *Polytechnisches Journal* 18 (1825): 484; "Literary and Scientific Intelligence," *La Belle Assemblée; or, Bell's Court and Fashionable Magazine* 10 (1825): 182; "Fixing of Crayon Colours," *Museum of Foreign Literature and Science* 8 (1826): 89; *Der bayerische Volksfreund* 3 (1826): 36; Charles Humphrys, *Methode zur Beseftigung der Farben von Pastell-Gemalden, Der englishe Zeichenmeister oder die neuesten Methoden* (Leipzig, 1832), 63.

237. "about practical science." *Mechanics' Magazine* 4 (1825): 379, 381; "James Smithson, A method of fixing crayon colours," *Franklin Journal and American Mechanics' Magazine* 1 (1826):

304. This was the journal of the Franklin Institute in Philadelphia. "Method of Fixing Crayon Colours," *Register of Arts and Journal of Patent Inventions* (London, 1828), 1:63–64; *Arcana of Science and Art; or One Thousand Popular Inventions and Improvements*, 3rd ed. (London, 1828), 211; and *Recreations in Science; or, a Complete Series of Rational Amusement* (London, 1830), 171.

237. "not as an artistic work." Smithson, "Crayon Colours," 236.

238. "not everyone viewed with approval." Birkbeck replaced Thomas Garnett as professor of Natural Philosophy at The Andersonian, in Glasgow, after Garnett took a position as chemistry lecturer at the Royal Institution in London. Smithson, as a founding member of the Royal Institution, likely heard about Birkbeck's activities from Garnett. Ewing, *Lost World*, 300; Thomas Kelly, *George Birkbeck: Pioneer of Adult Education* (Liverpool: Liverpool University Press, 1957); *The Chemist* 18 (July 17, 1824): 292.

238. "what was going on." John Godard, *George Birkbeck, the Pioneer of Popular Education* (London: Bemrose and Sons, 1884), 69, 72–74, 87.

239. "self-education of the English working class." The Society for the Diffusion of Useful Knowledge was founded by Henry Brougham in 1826, with the goal of diffusing practical knowledge throughout the British Empire.

239. "'Knowledge among men.'" William Rhees, *James Smithson and His Bequest* (Washington, DC, 1880), 24; emphasis added.

239. "some form of worker education." Bird has written intelligently about the phrase "increase and diffuse," although he failed to note the full significance of the Mechanics' Institute movement. William L. Bird Jr., "A Suggestion Concerning James Smithson's Concept of Increase and Diffusion," *Technology and Culture* 24 (1983): 246–255.

EPILOGUE: WHO WAS JAMES SMITHSON?

241. "he was in constant pain." This draws on Heather Ewing, *The Lost World of James Smithson* (New York: Bloomsbury, 2007), 271–342; Alan J. Rocke, *Nationalizing Science* (Cambridge, MA: MIT Press, 2001), 11–36.

242. "on his teeth on the left side." David R. Hunt, "James Smithson's Remains: A Biological Perspective," in *Engaging Smithsonian Objects through Science, History, and the Arts*, ed. Mary Jo Arnoldi (Washington, DC: * Institution Scholarly Press, 2016), 29–45.

243. "'admired his acquirements.'" Davies Gilbert, "Anniversary Address to the Royal Society, Nov. 30, 1829," *Philosophical Magazine* 7 (1830): 33–42.

243. "he seemed destined to be forgotten." Francois Beudant, *Traité élémentaire de minéralogie* (Paris, 1832), 2:354–357.

244. "someone's pet project." William J. Rhees, "James Smithson and His Bequest," *Smithsonian Miscellaneous Collections* 21 (Washington, DC, 1881), 2.

244. "play an important part." Walter R. Johnson, "A Memoir on the Scientific Character and Researches of James Smithson," *The Scientific Writings of James Smithson*, ed. William J. Rhees (Washington, DC, 1879), 123–142.

245. "accepted with little question." Ibid., 143–144.

245. "'extinct and forgotten.'" Rhees, "James Smithson and His Bequest," 2.

INDEX

Page numbers followed by *f* and *t* refer to figures and tables, respectively.